THE SYMPATHETIC MEDIUM

THE SYMPATHETIC MEDIUM

FEMININE
CHANNELING,
THE OCCULT, AND
COMMUNICATION TECHNOLOGIES,
1859–1919

JILL GALVAN

CORNELL UNIVERSITY PRESS
Ithaca and London

First published 2010 by Cornell University Press

Printed in the United States of America

Library of Congress Cataloging-in-Publication Data

Galvan, Jill Nicole, 1971–
  The sympathetic medium : feminine channeling, the occult, and communication technologies, 1859–1919 / Jill Galvan.
    p. cm.
  Includes bibliographical references and index.
  ISBN 978-0-8014-4801-0 (cloth : alk. paper)
  1. Channeling (Spiritualism)—History—19th century.
  2. Women mediums—History—19th century.
  3. Communication—Technological innovations—History—19th century.   4. Spiritualism in literature.
  5. Mediums in literature.   6. Communication in literature.   7. American literature—19th century—History and criticism.   8. English literature—19th century—History and criticism.   I. Title.

  BF1286.G35   2010
  133.9'1082—dc22

                                        2009028973

Cornell University Press strives to use environmentally responsible suppliers and materials to the fullest extent possible in the publishing of its books. Such materials include vegetable-based, low-VOC inks and acid-free papers that are recycled, totally chlorine-free, or partly composed of nonwood fibers. For further information, visit our website at www.cornellpress.cornell.edu.

Cloth printing   10 9 8 7 6 5 4 3 2 1

❧ Contents

Acknowledgments   *vii*

Introduction: Tuning in to
the Female Medium   1

1. Sympathy and the Spiriting
   of Information *In the Cage*   23

2. Securing the Line: Automatism and
   Cross-Cultural Encounters in
   Late Victorian Gothic Fiction   61

3. Du Maurier's Media: The Phonographic
   Unconscious on the Cusp of the Future   99

4. Telltale Typing, Hysterical Channeling:
   The Medium as Detective Device   135

5. Literary Transmission and
   Male Mediation   160

   Epilogue   188

Bibliography   *195*

Index   *209*

✍ ACKNOWLEDGMENTS

I have lived with this book a long time and owe many thanks to those who have enabled and guided my efforts along the way. I'm grateful to my graduate school mentor, Joe Bristow, for pushing me early on to clarify the terms of my argument and for continuing to be committed to my success in the profession ever since. Kate Hayles first motivated my interest in topics that would eventually lead me to this book and went on to offer invaluable early direction of the project. Many thanks also to Michael North and Mary Terrall for their insightful readings of this work in its earliest stages, and to Jonathan Grossman, who gave me excellent advice at a critical juncture. Over the years, many others have read sections or commented on aspects of the book, including David Brewer, Georgina Dodge, Richard Dutton, Linda Ferreira-Buckley, Beth Hewitt, Cricket Keating, Valerie Lee, Karen Leick, Marlene Longenecker, Jim Phelan, Elizabeth Renker, David Riede, Clare Simmons, Mytheli Sreenivas, Rebecca Wanzo, Roxann Wheeler, and members of UCLA's Nineteenth-Century Reading Group. I am also grateful to the very helpful anonymous readers at Cornell University Press, as well as to Peter Potter for offering editorial guidance that has been both incisive and encouraging.

Writing this book has given me the opportunity to engage in an exciting conversation with others about Victorian technologies, the occult, and gender. Joe Bristow, Jennifer Fleissner, Kate Hayles, Marty Hipsky, Laura Otis, and Anne Stiles all generously shared with me unpublished manuscript versions of their work. At numerous conferences, I have been challenged and enlightened by my fellow presenters and by audience members; I'm especially appreciative to Laura and many others who invited me to take part in their panels.

The rare-books collections at the Harry Ransom Center at the University of Texas at Austin were a great aid to my research. I also thank UCLA and its English Department and the Ohio State University's College of Humanities, which offered helpful financial support at different phases of the writing process. Earlier versions of material in chapters 1, 2, and 5 appeared as journal articles—respectively: "Class Ghosting 'In the Cage,'" *Henry James Review* 22 (2001): 297–306; "Christians, Infidels, and Women's Channeling in the Writings of Marie Corelli," *Victorian Literature and Culture* 31 (2003): 83–97; and "The Narrator as Medium in George Eliot's 'The Lifted Veil,'" *Victorian Studies* 48 (2006): 240–48. I am grateful to the editors for permission to use the material here.

Finally, I offer thanks to the people who have sustained me in many ways during the writing of this book. I have been inspired by the tenacity and optimism of my father, Carlos Galvan. My mother, Janice Galvan, gave much to enable my early successes and continues to be a constant source of affection and support. Last and most of all, I thank my husband, Dan Seward, for always listening and for being generous with his time, energy, and patience. This book is dedicated to you.

Introduction

Tuning in to the Female Medium

In *The Soul of Lilith* (1892) by Marie Corelli, a scientifically ambitious man uses a woman to discover the secrets of the heavens. The Lilith of the title lies entranced in a makeshift spiritual laboratory in a locked room in a London house, subjected to a brilliant Middle Easterner named El-Râmi. Having revived Lilith's dead body six years previously, El-Râmi now tethers her soul. While she is unconscious, her soul traverses the immortal regions of the universe, but he, a skilled metaphysical engineer, may call it back anytime he wishes, querying Lilith about her cosmic experiences. El-Râmi's ultimate goal: to glean empirical knowledge of God—His realm and whether He exists.

Lilith is a woman-turned-communication device, a way for El-Râmi to receive data from afar. Her body is a "machine" with what amounts to an on-off switch, and which he keeps running smoothly through regular injections of something called "Electro-flamma."[1] But if her function is techno-logical, it is also spiritual, in that the information El-Râmi seeks comes from a great beyond. That her mechanical body coupled with her soul-messages makes her so difficult to categorize—is she a personified telegraph, or more like a séance medium?—underscores that she represents not so much any

---

1. Marie Corelli, *The Soul of Lilith* (1892; London: Methuen, 1905), 37, 231.

one particular instrument or method as a communicative medial role. Lilith connects to another place, bringing new intelligence from there to here. At the same time, Corelli does not let us forget that this medium is a woman, emphasizing Lilith's beauty, which inspires a dangerous ardor in El-Râmi and a jealous sense of rivalry with his brother.

A bestselling author with a flair for crafting plots that resonated with the public, Corelli genders transmission in a way that would have been familiar to readers by the late Victorian period. The nineteenth century witnessed remarkable feats of transmission: ghostly messages turned up in séances, the telegraph and telephone sent words almost instantaneously through the "ether"—both created previously inconceivable forms of interpersonal connection. Uniting these communications was not just their mysteriousness but also their need for a go-between, someone to mediate them; very often it was a woman who carried out this task. *The Soul of Lilith* translates into the stuff of transcendental romance a figure that had emerged by Corelli's day in multiple locales, ranging from the office, a common site for women telegraphers and typists, and the telephone switchboard, an almost exclusively female workspace, to the séance, where women predominated as spirit and hypnotic channels.

Focusing on Britain and the United States around the late nineteenth and early twentieth centuries, this book explores depictions of the female communication go-between. As I argue, the work of mediating others' transmissions was a single vocation that took many specific forms—typing, telegraph operating, telephone operating, and occult mediumship; for reasons having to do with ideal conditions of dialogue and knowledge exchange, people viewed women as particularly well suited for this vocation. By examining this gendering, we gain a revealing vantage point onto nineteenth-century ideas of femininity as well as modernizing concepts of communication and knowledge transfer. One consequence of the feminization of channeling is that a distinct feminine trope arises in literature and culture. The female medium possesses certain recognizable emotional and psychological characteristics and is repeatedly implicated in situations involving private information and its antitheses, publicity and disclosure. Yet this figure is at once emblematic and pliable: she signifies diversely across individual narratives and spans multiple genres. *The Sympathetic Medium* tracks her pervasive and versatile literary presence.

Understanding the cultural place of the female medium requires appreciating the birth during the nineteenth century of a complex new world of technological and occult communications. Inventors came forward with a host of transmitting, transcribing, and recording technologies. In 1837,

Englishmen Charles Wheatstone and William Fothergill Cooke patented their needle version of the electric telegraph, while in the same year in the United States, Samuel Morse gave his first public demonstration of his armature telegraph. Guglielmo Marconi would make further strides on this front with his 1895 invention of the wireless (radio) telegraph. Photography likewise came into being in the 1830s; with the creation of the celluloid film roll in 1891, it became available for popular recreation. In 1873, E. Remington and Sons arranged with Christopher Latham Sholes to manufacture the typewriter, for which Sholes had received a patent five years previously. Alexander Graham Bell applied for a patent for the telephone in 1876 (as did, on the very same day, Elisha Gray), one year before Thomas Edison completed his phonograph.

Besides the innovations of photography and electric telegraphy, the 1830s saw the beginnings of an Anglo-American absorption with the paranormal (or questionably normal). Mesmeric lecturers and physicians from the Continent began showcasing their ability to manipulate an invisible animal magnetic fluid and thus to produce fantastic feats of mental and bodily "rapport" with their subjects. These included not only trance but also sometimes more spectacular effects in the magnetized "somnambulist"—a telepathic connection to the magnetizer, a clairvoyant ability to see faraway scenes or into the body's illnesses, even interaction with the spirit world. Though Franz Anton Mesmer had conceived of animal magnetism as a medicinal practice—a means to restore health by restoring a balance of vital fluid—it was not surprisingly its occult trappings that caught popular American attention, setting the stage for the séance craze. In 1848, young sisters Kate and Margaret Fox claimed to have inspired phantom knockings or "rappings" in their home in Hydesville, New York. Reports of the rappings stimulated a national fascination with spiritualism—a system of beliefs and practices for contacting the dead—which, a few years later, would wend its way to the British public. Devoted spiritualists took to the podium to spread the word of their discoveries; in the decades that followed, many others held séances featuring table-tipping, the materialization of spirits, and planchette and automatic writing.[2]

---

2. Adam Crabtree, *From Mesmer to Freud: Magnetic Sleep and the Roots of Psychological Healing* (New Haven: Yale University Press, 1993); Alison Winter, *Mesmerized: Powers of Mind in Victorian Britain* (Chicago: University of Chicago Press, 1998); Daniel Pick, *Svengali's Web: The Alien Enchanter in Modern Culture* (New Haven: Yale University Press, 2000); Slater Brown, *The Heyday of Spiritualism* (New York: Hawthorn Books, 1970); Janet Oppenheim, *The Other World: Spiritualism and Psychical Research in England, 1850–1914* (Cambridge: Cambridge University Press, 1985).

In 1882, a group of Cambridge University scholars founded the Society for Psychical Research (SPR), an organization committed to the scientific research of mesmerism, telepathy, and other extraordinary or otherworldly phenomena. In strict investigations, psychical researchers staked their claim by their empiricism, though for some, like Frederic W. H. Myers, this was mixed with an underlying hope in the existence of a spiritual plane of human activity. Indeed, these psychical researchers, like some spiritualists, depended on forays into the occult as a substitute for religious faith—as a means of recovering, in the Victorian period of increasing disbelief, evidence of an immortal soul.[3] Others mostly valued the SPR for its researches into unconscious mental faculties, which psychical researchers often concluded were, in one way or another, what lay behind the bizarre occurrences they studied. Psychologists sometimes crossed paths with psychical researchers in their attempts to understand the hypnotic trance and other effects of the subterranean mind. The SPR also developed a branch in the United States and soon boasted a number of famous members, including explorer Henry Stanley, authors Arthur Conan Doyle and Lewis Carroll, MP Arthur Balfour, criminologist Cesare Lombroso, philosopher William James, and scientists William Crookes and Oliver Lodge. Prime Minister William Gladstone, also a member, probably voiced a common opinion within the SPR when he called its research "the most important work . . . being done in the world. By far the most important."[4]

The Fox sisters' catalyzing of "modern spiritualism" brings home the crucial role that girls and women played in the movement. On both sides of the Atlantic, spirit channeling quickly became marked as feminine.[5] What is striking, and what has been the impetus of this book, is that the development of female mediumship parallels women's increasing involvement over the course of the period in technological modes of communication mediation. Western Union, which would soon become a monopoly in the United States, hired its first female telegrapher in 1846. Though the number of nineteenth-century

---

3. On the spiritual consolations of occultist research, see Oppenheim, *Other World*.

4. Quoted in R. Laurence Moore, *In Search of White Crows: Spiritualism, Parapsychology, and American Culture* (New York: Oxford University Press, 1977), 139. On the SPR's illustrious member list, see Oppenheim, *Other World*, 135, 141, 245, and 373.

5. Accounts of women and Victorian spiritualism note that ideal feminine characteristics dovetailed with the requirements of mediumship: passivity (to the will of a foreign personality), sensitivity (to the elusive nature of that will), and a strong spiritual sensibility. See Ann Braude, *Radical Spirits: Spiritualism and Women's Rights in Nineteenth-Century America* (Boston: Beacon, 1989), 23–24; Alex Owen, *The Darkened Room: Women, Power, and Spiritualism in Late Victorian England* (Philadelphia: University of Pennsylvania Press, 1990), 7–10, 28ff., 63; Moore, *In Search of White Crows*, 105–6; and Jennifer A. Yeager, "Opportunities and Limitations: Female Spiritual Practice in Nineteenth-Century America," *ATQ* 7, no. 3 (1993): 217.

female telegraphers was never large in absolute terms, its continued increase is itself notable: by 1900, women had garnered 13 percent of U.S. telegraphy positions.[6] The size of the female telegraphic workforce rose more rapidly in Britain, reaching over a quarter of all spots by 1880 at the Post Office, which had taken over the telegraph as a public utility a decade earlier.[7] In the early twentieth century, telegraphy emerged as a solidly gendered field among users of teletype,[8] because by then typing itself had crossed over into the category of feminine employment. In the United States, women represented 64 percent of stenographer-typists in 1890, then virtually all of them thirty years later.[9] Britain also tended toward the feminization of this job sector: from 1851 to 1911, the number of male clerks grew sevenfold, while that of female clerks increased by a stunning factor of eighty-three, or from 2 percent to 20 percent of the overall clerical workforce; in the commercial industries, women were especially linked to the mushrooming fields of typing and shorthand.[10] They made even speedier inroads as telephone operators. By the late 1880s, when Bell's invention was just a decade old, they constituted most of the daytime workforce in both Britain and the United States. The British system would continue well into the 1900s to hire men for nighttime operating, but U.S. industry leader Bell Telephone ceded even those posts to female workers in 1904, to become the largest early twentieth-century employer of women.[11]

---

6. Annteresa Lubrano, *The Telegraph: How Technology Innovation Caused Social Change* (New York: Garland, 1997), 133; Thomas C. Jepsen, *My Sisters Telegraphic: Women in the Telegraph Office, 1846–1950* (Athens: Ohio University Press, 2000), 59. But Jepsen notes that U.S. census figures probably undercount the actual number of women operators; popular literature and telegraph journals indicate the existence of many more. According to the estimates of a labor leader and Western Union executive questioned during an 1883 strike, women at that time already made up 20–25 percent of telegraph operators (53–54).

7. Jeffrey Kieve, *The Electric Telegraph: A Social and Economic History* (Newton Abbot, UK: David and Charles, 1973), 190.

8. Jepsen, *My Sisters Telegraphic*, 8, 11, 34. Instead of Morse code, teletype used a combination of typewriter, data code, and punched tape to transmit messages.

9. Margery W. Davies, *Woman's Place Is at the Typewriter: Office Work and Office Workers, 1870–1930* (Philadelphia: Temple University Press, 1982), appendix, table 1.

10. Meta Zimmeck, "Jobs for the Girls: The Expansion of Clerical Work for Women, 1850–1914," in *Unequal Opportunities: Women's Employment in England 1800–1918*, ed. Angela V. John (Oxford: Blackwell, 1986), 154; Gregory Anderson, *Victorian Clerks* (Manchester: Manchester University Press, 1976), 60, 103. Samuel Cohn defines Victorian "clerks" as synonymous with office workers but explains that the job title was more comprehensive than is true today; *The Process of Occupational Sex-Typing: The Feminization of Clerical Labor in Great Britain* (Philadelphia: Temple University Press, 1985), 33–34, 68. His and Zimmeck's analyses make clear that clerical work ranged from word processing, mail sorting, and bookkeeping to telegraphy and telephony.

11. Peter Young, *Person to Person: The International Impact of the Telephone* (Cambridge: Granta, 1991), 26; Brenda Maddox, "Women and the Switchboard," in *The Social Impact of the Telephone*, ed. Ithiel de Sola Pool (Cambridge, MA: MIT Press, 1976), 276; Stephen H. Norwood, *Labor's Flaming Youth: Telephone Operators and Worker Militancy, 1878–1923* (Urbana: University of Illinois Press, 1990), 4, 28.

Concerns about labor availability and management played a part in the feminization of these technological fields. Historical developments created supply problems that women helped to solve; by the late nineteenth century, for instance, the growth of capitalist industry and government bureaucracies created an increasing need for typists.[12] Employers may have also leaned toward the sex they saw as more "docile" and therefore less likely to be involved in labor disputes.[13] In addition, they saved money by hiring women because women in general accepted lower wages than those demanded by men.[14] Focusing on the lower wages of female operators and typists, however, begs the question of why feminization was not more common in other occupations.[15] More broadly and significantly, that question suggests the danger of trying to theorize feminization by viewing operating or typing only as abstract labor—as a commodity to be found and bought or as a force to be contained: we are left wondering why we could not substitute any type of labor for the one under consideration. Comprehending more fully why this work went to women entails paying more attention to its specific nature. At the same time, there is much value in thinking about feminization not simply as a negative choice—as a matter of *not* choosing workers who would be too unwieldy or too expensive—but as a positive choice, a choice for workers believed capable of outperforming others in the particulars of the job. That is, it is logical to analyze feminization in terms of femininity.[16]

---

12. See Davies, *Woman's Place*, as well as her "Women Clerical Workers and the Typewriter: The Writing Machine," in *Technology and Women's Voices: Keeping in Touch*, ed. Cheris Kramarae, 29–40 (New York: Routledge and Kegan Paul, 1988). In another argument turning on labor-supply pressures, Thomas C. Jepsen suggests that men's military service in the American Civil War opened up jobs for women in Western frontier telegraph; "Women Telegraph Operators on the Western Frontier," *Journal of the West* 35 (1996): 75.

13. See for example Cohn on the hiring of women as telegraphers in the British Post Office (*Process of Occupational Sex-Typing*, 123, 190). Notably, Cohn's own finding is that gender difference in and of itself ultimately played a negligible role in clerical workers' labor militancy (chap. 7).

14. See Cohn, *Process of Occupational Sex-Typing*; Kieve, *Electric Telegraph*, 190; and Maddox, "Women and the Switchboard," 266. A related theory is that clerical work de-skilled beginning in the nineteenth century and that this resulted in feminization because, as Cohn summarizes the argument, "historically, women have been confined to low-status occupations characterized by low levels of responsibility, inferior rates of pay, and limited prospects for promotion... the increasing undesirability of office jobs would have facilitated the use of women for these positions" (65). But Cohn finds the de-skilling claim questionable in the case of clerical work, noting that typing and shorthand actually required more skill than the hand-copying and other methods that preceded them (66–68, 84–85). Telegraphy also entailed substantial training and skill (123).

15. Several scholars have noted this fault in the wages argument; for example, see Davies on typing in *Woman's Place*, 56.

16. This kind of analysis takes into account women's reputation for finer dexterity with technologies like the typewriter; see Sally Mitchell, *The New Girl: Girls' Culture in England, 1880–1915* (New York: Columbia University Press, 1995), 35. But as Mitchell suggests, even as applied today,

Indeed, Victorian and turn-of-the-century perceptions of women in communication mediation prompt us to consider gender more closely in at least a couple of respects. First, the discourse around these workers frames them in iconic terms that point up their identity as "girls" or young women. The telephone operator was the customer's "hello-girl." Fictional accounts of typists and telegraphers often underscore a heroine's femininity within the very title, as we see in Anthony Trollope's "The Telegraph Girl" (1877), Grant Allen's *The Type-Writer Girl* (1897), and Tom Gallon's *The Girl Behind the Keys* (1903). In Allen's novel, typing is a distinctly feminine pursuit: "every girl in London can write shorthand, and . . . type-writing as an accomplishment is as diffused as the piano."[17] And the earliest advertisements for and demonstrations of the typewriter featured women at the machine.[18]

Second, mediating jobs enjoyed a respectability at a time when women's social and gender status were closely intertwined. At the turn of the century, "girls"—roughly speaking, females beyond school age but still unmarried—were considered to be within a liminal period in which paid work was generally socially acceptable; nonetheless, some girls' jobs were more securely gendered and concomitantly of a higher status than others.[19] Billed as white-collar work, typing and switchboard operating were attractive options for young women who needed employment but also wanted to remain "ladies." Due in part to the fact that it involved mixed company, female telegraphy occupied a somewhat lower social stratum; even still, it drew the daughters of clergymen, tradesmen, and other middle-class men in Great Britain and, likewise, allowed the American woman wage-earner to maintain a degree of gentility.[20] Certain inherent features such as the prerequisite of a general or specialized education no doubt contributed to the social standing of communications work. Yet it is worth pressing harder on the question of why mediating jobs would have been more or less exceptional in being compatible

such a rationale seems limited once we consider professions (brain surgery, for example) requiring great dexterity yet dominated by men (*New Girl*, 193 n. 34).

17. Grant Allen [Olive Pratt Rayner, pseud.], *The Type-Writer Girl* (1897; Peterborough, Ont.: Broadview, 2004), 28.

18. Carole Srole, "'A Blessing to Mankind, and Especially to Womankind': The Typewriter and the Feminization of Clerical Work, Boston, 1860–1920," in *Women, Work, and Technology: Transformations,* ed. Barbara Drygulski Wright et al. (Ann Arbor: University of Michigan Press, 1987), 87.

19. On working and "girlhood," see Mitchell, *New Girl;* chap. 2 discusses girls' work and notes typing and telephony's distinction as decorous feminine employment (35).

20. See Edwin Gabler, *The American Telegrapher: A Social History, 1860–1900* (New Brunswick, NJ: Rutgers University Press, 1988), 125; though Gabler notes that the working-class families of the telegraphers made their gentility "peculiarly wobbly" (128). On British women telegraphers and class backgrounds, see Kieve, *Electric Telegraph,* 85.

with bourgeois notions of femininity. Just as we understand the relative gentility of Victorian governesses to result from their public application of reputedly womanly (in that case, maternal) tendencies, it makes sense to ask what traits were presumed to enter into typing/shorthand, telephony, and telegraphy, fueling their reputation as proper women's work.

We also need to take note of the particular character of these jobs and what they have in common with one another. That a nineteenth-century wave of feminization mainly encompassed the work of conveying dialogue or knowledge from one person or site to another certainly invites further analysis. Compounding the significance of this trend is the fact that a shadow version of the same activity arose and was gendered female among occultists. Moreover, there are certain key similarities in how the women performing each of these mediating functions were envisioned by their contemporaries. These similarities suggest that focusing on the feminization of, say, spirit mediumship or typing in isolation risks missing a bigger picture.

Although it may seem odd to group séance behavior, typing, and operating under one vocational rubric, writings of the period encourage this composite view, in that they often associate occult modes of communication and projection with technological ones. New machines for talking to distant selves evoked ideas of haunting. One contributor to a late nineteenth-century periodical imagined the humming noises that corrupted the telephonic connection as the wailing of wandering spirits.[21] Another early user compared the experience of listening to a friend over the wire to hearing a "voice from another world. Here I was speaking to a person far away from me whom I could hear as though he was at my side and yet I could not see him."[22] Inversely and more frequently, those interested in paranormal contacts took new communication devices as their models and aids.[23] The origin of modern spiritualism

---

21. Young, *Person to Person,* 32.

22. Quoted in Claude S. Fischer, *America Calling: A Social History of the Telephone to 1940* (Berkeley: University of California Press, 1992), 62.

23. Much recent work has helped to chart these links, including Marina Warner, *Phantasmagoria: Spirit Visions, Metaphors, and Media into the Twenty-first Century* (Oxford: Oxford University Press, 2006); Pamela Thurschwell, *Literature, Technology and Magical Thinking, 1880–1920* (Cambridge: Cambridge University Press, 2001); Jeffrey Sconce, *Haunted Media: Electronic Presence from Telegraphy to Television* (Durham: Duke University Press, 2000); Steven Connor, "The Machine in the Ghost: Spiritualism, Technology and the 'Direct Voice,'" in *Ghosts: Deconstruction, Psychoanalysis, History,* ed. Peter Buse and Andrew Stott, 203–25 (New York: St. Martin's Press, 1999); Lawrence Rainey, "Taking Dictation: Collage Poetics, Pathology, and Poetics," *Modernism/Modernity* 5, no. 2 (1998): 123–53; Allen W. Grove, "Röntgen's Ghosts: Photography, X-Rays, and the Victorian Imagination," *Literature and Medicine* 16 (1997): 141–73; Tom Gunning, "Phantom Images and Modern Manifestations: Spirit Photography, Magic Theater, Trick Films, and Photography's Uncanny," in *Fugitive Images: From Photography to Video,* ed. Patrice Petro, 42–71 (Bloomington: Indiana University Press, 1995);

was, famously, a harbinger of this trend. The Fox sisters conversed with their ghost by means of a code that matched up each letter of the alphabet with a specific number of rappings (one meant "A," two meant "B," and so on), and it might have been this percussive linguistic code that prompted the labeling of the rappings as a "spiritual telegraph" and the report that "God's telegraph has outdone Morse's altogether!"[24] The extramundane shaded into the technological in literary works by all kinds of authors, including some we regard as canonical. In Charles Dickens's 1858 "The Rapping Spirits," a facetious take on spiritualism, a boy shams mediumship and jerks his limbs around as if possessed, "conduct[ing] himself like a telegraph before the invention of the electric one."[25] Eight years later, Dickens adopted a graver tone in another Christmas story, "The Signal Man," which juxtaposes telegraphic warnings with ghostly ones, as a railway signalman, charged with ferrying messages that control traffic, struggles to interpret an apparition that seems to herald an accident.

But comparisons of the occult and technology amounted to more than the convenient conceits of commentators and fiction writers. Or to be more precise, even these conceits were not necessarily idle but rather sometimes sprang from a sincere interest in exploring resemblances and interrelationships. Occultists often spoke of machines permitting remote conversation through invisible processes as only some distant and telling order of their own brand of communication. Calling telephony and radio telegraphy "valuable as enabling us to illustrate the difficulties as well as the possibilities of proving the existence of life after death," journalist and spiritualist W. T. Stead posed as the ultimate question whether automatic writing (an already proven instrument of telepathy, in his opinion) could "be extended to those who have crossed the river of death—an extension which corresponds to the transmission of Marconigrams across the Atlantic."[26] Stead's 1897 writing recalls mid-nineteenth-century attempts to make spiritualism seem empirically plausible: our use of the term *supernatural* to describe séance-goers' pursuits misrepresents their own convictions that spiritualistic communications came down to ultimately knowable natural laws rooted in scientifically established properties, ones for example akin to electromagnetism.[27]

---

and Avital Ronell, *The Telephone Book: Technology, Schizophrenia, Electric Speech* (Lincoln: University of Nebraska Press, 1989).

24. Quoted in Brown, *Heyday of Spiritualism,* 110.

25. Charles Dickens, "The Rapping Spirits," in *Charles Dickens' Christmas Ghost Stories,* ed. Peter Haining (New York: St. Martin's Press, 1993), 199.

26. W. T. Stead, *After Death: A Personal Narrative* (1897; London: Review of Reviews, 1914), x, xvii.

27. Richard Noakes describes spiritualism's purported natural laws in "Spiritualism, Science and the Supernatural in Mid-Victorian Britain," in *The Victorian Supernatural,* ed. Nicola Bown, Carolyn

Some would take a similar approach to theories of telepathy. William Walker Atkinson's *Practical Mind-Reading* (1908) quotes French astronomer Camille Flammarion on the quasi-telephonic transfer of thought between individuals: "The transformation of a psychic action into an ethereal movement, and the reverse, may be analogous to what takes place on a telephone, where the receptive plate, which is identical with the plate at the other end, reconstructs the sonorous movement transmitted, not by means of sound, but by electricity."[28] For Flammarion, electricity is only a metaphor for ether, converting and carrying sound in the same way that ether does psychic content; but for others, electricity was the actual agent of psychical communication. In 1892, William Crookes, a renowned British physicist and chemist, speculated that in "some parts of the human brain may lurk an organ capable of transmitting and receiving other electrical rays of wave-lengths hitherto undetected by instrumental means. . . . In such a way the recognised cases of thought transference, and the many instances of 'coincidence' would be explicable."[29] Crookes had shown an ardent interest in spiritualism in the 1870s, when he joined up with electrician and telegraph engineer Cromwell Fleetwood Varley to prove the honesty of celebrated medium Florence Cook by connecting her body to a mild galvanic circuit. The circuit was designed to prevent Cook, who began her séances within a curtained cabinet, from impersonating the spirit who materialized from it, since even the faintest bodily movements would alter the readings of the galvanometer (one of Varley's telegraph cable-testing devices) stationed just outside the cabinet.[30]

While this test (which remarkably turned up no sign of trickery) adjoined the séance body to an electrical circuit in an artificial way, modern spiritualists also professed séance bodies' natural connections to electricity and inherited mesmerists' tendency for magnetic explanations of their practices. In 1779, Franz Anton Mesmer, father of animal magnetism, met with Benjamin Franklin to talk up his discovery, but the commission Franklin chaired to investigate it ultimately decided against its validity.[31] In the mid-nineteenth century, though, Franklin reemerged, this time in spirit form, and considerably more amenable to the idea of invisible influences, to announce that

Burdett, and Pamela Thurschwell, 23–43 (Cambridge: Cambridge University Press, 2004). See also Moore, *In Search of White Crows*, 7–36 passim.

28. William Walker Atkinson, *Practical Mind-Reading* (Chicago: Lyal Book Concern, 1908), 10.

29. William Crookes, "Some Possibilities of Electricity," *Fortnightly Review* 51 (1892): 176.

30. Richard J. Noakes, "Telegraphy is an Occult Art: Cromwell Fleetwood Varley and the Diffusion of Electricity to the Other World," *British Journal of the History of Science* 32 (1999): 450–58; Oppenheim, *Other World*, 345–46.

31. Crabtree, *From Mesmer to Freud*, 23–29, 213.

spiritualism had been made possible by his own electrical findings. More so than any other religious, philosophical, or national personality, Franklin maintained a presence during these decades in the visions and writings of the American spiritualist community, sometimes giving lessons and demonstrations on electricity and magnetism, especially their capacity to promote spirit contacts.[32] Séance sitters sometimes joined hands in a circle they dubbed a "circuit" or "battery" inasmuch as it put them in technological harmony with the spirit world.[33] Alternating men (conceived as positively or actively charged) with women (conceived as negative or passive) enhanced the circle's usefulness as a station for point-to-point dialogue.[34] Andrew Jackson Davis, a Swedenborgian and early spiritualist, counseled that sitters should hold onto a "magnetic cord" for about the first hour; his writings depicted this arrangement with a diagram similar to those illustrating Victorian textbooks on electricity and magnetism.[35]

Given this context, Jeffrey Sconce hardly exaggerates in calling spirit mediums "wholly realized cybernetic beings—electromagnetic devices bridging flesh and spirit, body and machine, material reality and electronic space."[36] In the cultural imagination, spirit mediums were much closer to other mediating women hooked up to wires or machines—the operator at her switchboard, the telegrapher at her sounder, the secretary at her typewriter—than is at first recognizable.[37] They were similarly implicated in technological networks, and, equally importantly, they participated in similarly structured acts of communication. It is particularly with this in mind that this book reads women's operating, typing, and séance channeling not as separate functions but as different expressions of the same one. Ideas of femininity spoke to aims and concerns about the human-relayed transfer of conversation and knowledge per se. These views underwrote women's welcome into

32. Moore, *In Search of White Crows*, 20; Werner Sollors, "Dr. Benjamin Franklin's Celestial Telegraph, or Indian Blessings to Gas-Lit American Drawing Rooms," *American Quarterly* 35 (1983): 459–80.

33. Bret E. Carroll, *Spiritualism in Antebellum America* (Bloomington: Indiana University Press, 1997), 132; Crabtree, *From Mesmer to Freud*, 247.

34. Carroll, *Spiritualism in Antebellum America*, 135; Braude, *Radical Spirits*, 23–24.

35. Carroll, *Spiritualism in Antebellum America*, 135–37. On British spiritualists' claims of electrical and magnetic properties in the séance, see Richard Noakes, "'Instruments to Lay Hold of Spirits': Technologizing the Bodies of Victorian Spiritualism," in *Bodies/Machines*, ed. Iwan Rhys Morus (Oxford: Berg, 2002), 129, 149.

36. Sconce, *Haunted Media*, 27.

37. Moreover, the sense that mediums functioned, as did women in offices or at switchboards, within networks is strengthened by the fact that they often communicated with the other side through spiritual intermediaries, known as controls.

mediating posts and helped to accustom the public to that occurrence, reinforcing their vocational prevalence.

The female medium arose alongside developments in the way people were able to communicate with one another. I argue that with the advent of both the transatlantic spiritualism craze and several communication technologies—some of them quite sensational—human-mediated exchange became especially visible as a category of communication. This category had of course existed before, in the form of hand-copying, which Victorian fiction continued to represent in characters like Dickens's Nemo or Melville's Bartleby. But mediation gained particular prominence and complexity over the course of the nineteenth century: the proliferation of methods produced a networked culture that offered dramatic new communicative opportunities as well as challenges, ones to which the female sex seemed especially well poised to respond. For women made at-a-distance dialogues seem less of what they were—less distanced and less burdened by the existence of an intermediary. Portraits of female media of all kinds commonly return to two allegedly feminine traits: sensitivity or sympathy, often imagined as the product of women's delicate nervous systems; and an easy reversion to automatism, or a state of unconsciousness. While the first posited the medium's ability to reach out feelingly to others and thus to facilitate networks of communication, the second presumed that such self-extension would *only* be a matter of feeling: by subtracting her intellectually from the path of communication, automatism eased and protected others' dialogue.

In short, women were exemplary go-betweens because they potentially combined the right kind of presence with the right kind of absence. The modernizing nineteenth century was marked by separations between individuals: physical and bureaucratic ones, demonstrated by the very technologies needed to span them; and existential ones—separations from family and friends brought on by death and which a swelling voice of evolutionists and other scientific discoverers, steeped in materialism and doubts about the afterlife, pronounced as final. Yet as presence, the mediating woman functioned as an important component in the sometimes practical, sometimes comforting premise that these separations were not absolute, as she helped to connect people despite obstacles of circumstance, space, and even mortality.

On the other hand, as absence, she allayed worries about privacy, for all human media theoretically stood in positions to learn or disseminate the knowledge that came their way. Other critics have remarked the frequency of the "automatic" label on typists and spirit mediums in particular, stressing its consequences for issues of textual authority or its usefulness for preserving an ideal of naive, untouched femininity. However, there has been little focused study of the gendering of the label together with what this gendering can tell us about cultural attitudes regarding communication and knowledge transfer. Notably, many U.S. states created laws punishing employees who disclosed the content of telegraph messages or telephone conversations.[38] Moreover, the nineteenth century witnessed the development of people's notions of privacy and vulnerable knowledge as such. Alexander Welsh argues that the British penny post and the telegraph, with their actually or potentially surveillant human agents, triggered worries about privacy, even as the growing newspaper press, an organ of public exposure, solidified the conceptual divide between privacy and publicity. Concomitantly, people placed an increasing premium on information, which Welsh defines as knowledge that can be stored for future use and used by people other than its original possessor. Information is "potentially exchangeable," a social commodity, deriving its greatest value through concealment (as in blackmail) or revelation (as in patents).[39] Human-relayed communication rendered more conspicuous, more inescapably a reality, the activity of exchanging words and knowledge, and I suggest that auxiliary individuals' possible interest in this information as it traveled via telecommunication networks resulted in a gendered paradigm of the disinterested conduit. The link between the privacy of information and the female medium's inattentiveness was as important in séance networks as in technological ones, though more indirectly so: because her automatism implied her ignorance of the often personal and confidential facts she communicated from the departed, it also implied the authenticity of the communications themselves, thereby verifying spiritualism.

I have chosen the term female *medium* to deliberately conjure up (by aid of modern connotations) an image of machinery. The women in the stories this book examines supplement or substitute for machines, often based on the premise of their mental automatism. Also implicit in this metaphor is the woman medium's manipulability: in several works, she is not an individual in herself but rather the tool of another's, usually a man's, design of gaining information or social power. This was the gender dynamic that obtained in

---

38. Paul Starr, *The Creation of the Media: Political Origins of Modern Communications* (New York: Basic Books, 2004), 187; Ithiel de Sola Pool, *Forecasting the Telephone: A Retrospective Technology Assessment of the Telephone* (Norwood, NJ: Ablex, 1983), 140. The premium on privacy is exemplified in Bell System executives' worry that switchboard operators' "breaches of confidentiality would 'seriously retard the increase of the toll business'" and in the emphasis in operating manuals on the need for secrecy; quoted in Venus Green, *Race on the Line: Gender, Labor, and Technology in the Bell System, 1880–1980* (Durham: Duke University Press, 2001), 51.

39. Alexander Welsh, *George Eliot and Blackmail* (Cambridge, MA: Harvard University Press, 1985), 43.

offices and other telecommunication sites, where female workers carried out the business of male managers; it also characterized the relationship between mediums and psychical researchers.[40] Of course, at bottom, the medium's unthinking compliance was probably always only theoretical. In actuality, typists petitioned for better workplace treatment, and operators went on strike and otherwise disobeyed their bosses, just as spirit mediums could be caught cheating.[41] Overtly or covertly, these women all acted purposefully and in their own self-interest. The female medium's failure to be an indifferent apparatus likewise generates key moments of drama and character development in some narratives under analysis here. Yet the very tensions these circumstances produced clues us into the sense of instrumentality essential to perceptions of the female medium. I am as interested in recognizing the disturbed paradigm as I am in interpreting the implications of the disturbances themselves. One facet of this book is indeed that it explores some of the manifold ways literature incorporates the concept of female automatism. Tim Armstrong, Lisa Gitelman, and others have shown us the importance of notions of mechanically inattentive or distracted writing at the turn of the twentieth century, but there has yet to be a broad exploration of narrative engagements with such concepts—for example, conflicting attitudes toward them—as they revolve around notions of femininity.[42]

This study is thus significantly attuned to automatism, but to say merely that female media were imagined as mechanical overlooks the full range of virtues they were supposed to bring to communication scenarios. Analyzing the impact of new technologies on concepts of inscription around 1900, influential critic Friedrich A. Kittler notably denies that gender mattered

---

40. See Vieda Skultans, "Mediums, Controls, and Eminent Men," in *Women's Religious Experience: Cross-Cultural Perspectives,* ed. Pat Holden, 15–26 (London: Croom Helm, 1983).

41. On typists' petitions, see Zimmeck, "Jobs for the Girls," 168–70. Morag Shiach finds that many fictions about typists engage with the issue of whether or not the profession can lead them to self-realization and erotic fulfillment; *Modernism, Labour and Selfhood in British Literature and Culture, 1890–1930* (Cambridge: Cambridge University Press, 2004), 70–78. On operators' disobedience, see for example Michèle Martin, *"Hello, Central?": Gender, Technology, and Culture in the Formation of Telephone Systems* (Montreal: McGill-Queen's University Press, 1991), 104–9. Eusapia Palladino was just one of many discovered fraudulent mediums; on her, see Ruth Brandon, *The Spiritualists: The Passion of the Occult in the Nineteenth and Twentieth Centuries* (New York: Knopf, 1983), 129, 138.

42. See Tim Armstrong on "Distracted Writing" (chap. 7) in *Modernism, Technology and the Body: A Cultural Study* (Cambridge: Cambridge University Press, 1998), and Lisa Gitelman on "Automatic Writing" (chap. 5) in *Scripts, Grooves, and Writing Machines: Representing Technology in the Edison Era* (Stanford: Stanford University Press, 1999). While there has been a scarcity of extended literary studies of female automatism, there have been explorations of female automata, wherein the emphasis falls on ideas of artifice; see for example Felicia Miller Frank, *The Mechanical Song: Women, Voice, and the Artificial in Nineteenth-Century French Narrative* (Stanford: Stanford University Press, 1995).

at all, declaring that the typewriter made for the "desexualization of writing" in permitting women entrance into the realm of textual production.[43] Conflating functionality with agency, this claim is problematic in and of itself. But in addition, there is a failure to recognize how women might be seen to invest communication with something beyond the mechanical or material, a misperception partly founded on an underlying notion that discursive technologies and inspired discourse were mutually exclusive at the turn of the century. In 1800, Kittler says, handwriting communicated the mind to the page through a seamless interconnection of body (hand and eye) and spirit, while by 1900 typing had severed that connection. There was no mind, in the Cartesian sense, in the discourse of 1900; what passed for meaning in typescript was the product of symbolic differences, and the figure at the typewriter was only its bodily complement, a neurally, psychophysically precise operator.[44]

In short, Kittler repudiates the possibility that late Victorian apparatus could seem to communicate the non-bodily—human intention and fundamental selfhood: at the limit, a soul—and therein facilitate genuine personal understanding between senders and recipients.[45] This disavowal of communicating subjectivities goes hand in hand with Kittler's underreporting of the persistent interest around 1900 in issues of spirit and the occult.[46]

---

43. Friedrich A. Kittler, *Gramophone, Film, Typewriter*, trans. Geoffrey Winthrop-Young and Michael Wutz (Stanford: Stanford University Press, 1999), 187.

44. Compare Mark Seltzer, who takes the typist as described by Kittler as an instance of the "body-machine complex," a motif of the coming together of the natural and the technological that he suggests proliferates in naturalist and realist texts; *Bodies and Machines* (New York: Routledge, 1992), 10–11. See also Bernhard Siegert, who contrasts Romantic and modern discourse in ways highly similar to Kittler's, though focusing not on media technologies but instead on evolving postal methods of discourse transmission; *Relays: Literature as an Epoch of the Postal System*, trans. Kevin Repp (Stanford: Stanford University Press, 1999).

45. Kittler makes this repudiation explicit in his discussion of the rise of modern media—from the typewriter to film—in tandem with neurological theories of mind: "The hard science of physiology did away with the psychological conception that guaranteed humans that they could find their souls through handwriting and rereading. . . . The unity of apperception disintegrated into a large number of subroutines, which, as such, physiologists could localize in different centers of the brain and engineers could reconstruct in multiple machines. Which is what the 'spirit'—the unsimulable center of 'man'—denied by its very definition" (*Gramophone, Film, Typewriter*, 188). Jay Clayton has incisively argued that Kittler discounts communication itself as a function of modern media and that he barely considers patently communicative devices like the telegraph and telephone because they disturb his account; *Charles Dickens in Cyberspace: The Afterlife of the Nineteenth Century in Postmodern Culture* (Oxford: Oxford University Press, 2003), 65–70. That denial of communication, I would also note, correlates with his denial of essential and intentional selfhood, since communication presumes thoughtful subjectivities who transfer their thoughts to one another.

46. Kittler does note spiritualistic applications of the telegraph and camera, but quite briefly (*Gramophone, Film, Typewriter*, 11–13); moreover, these mentions cast the technological manifestation

Psychophysics and psychoanalysis receive great attention in his *Discourse Networks, 1800/1900,* while psychical research is dismissed in a mere phrase: "*Ecriture automatique* appeared as early as 1850, but only among American spiritualists; it was not analyzed until the turn of the century. After the theoretical work of F. W. H Myers and William James, profane automatic writing arrived in the Harvard laboratory of the German psychologist and inventor of psychotechnology Hugo Münsterberg." That statement gives the anachronistic impression that the "theoretical work" of SPR men like Myers and James was over almost before it started and that automatic writing was swiftly removed from the realm of occultism to become the object of "profane" psychological investigations alone. Kittler concludes that within such experiments, "automatic writing says nothing of thought or inwardness, of intention or understanding; it speaks only of speech and glibness," ignoring the dreams of higher understanding and soulful exchange of many interested in automatic writing at this time, from psychical researchers to séance mourners of the dead soldiers of the world wars.[47]

Acknowledging the actual ongoing interest in occultist automatic writing and, moreover, how often the occult informed visions of technologies for writing or communicating can help us to realize the desire for authentic interconnection—the transfer, precisely, of thought or inwardness, intention or understanding—through bodily relays of various kinds. My account insists on the prospect of communicating one's meaning and one's self in discussions of turn-of-the-century communication machinery. I view this possibility as a specific consequence of the gender of the body at the machine. For feminine nerves and temperaments were seen as crucially distinct from men's: only women generated what we might phrase as a sympathetic excess—an affective or spiritual quality—that could transform mediating apparatus into the carriers of intentional self-to-self communication. This was most vividly the case at the séance, where the medium's sensitivity supplemented electromagnetic arrangements to enable personal contact between worlds, but we can also see it in more subtle form within mundane settings: for example, in the operator who patiently and considerately connected others through the switchboard or in the secretary who knew her employer's volitions so well that she could stand for him in business interactions. This excess and its role in transmitting subjectivity have escaped most accounts of the automatized

---

of the dead as an effusion along the lines of the Lacanian real and thus treat somewhat flippantly spiritualists' serious hopes of communicating with recognizable selves and gleaning divine meaning.

47. Friedrich A. Kittler, *Discourse Networks, 1800/1900,* trans. Michael Metteer, with Chris Cullens (Stanford: Stanford University Press, 1990), 225, 228.

female producer of turn-of-the-century discourse.[48] To notions of mechanical utility, we must add perceptions of a feeling and moral aptitude in women for establishing interpersonal networks and appreciate how expressions of mediumistic sympathy can become the focus of narrative.

The mediating woman is distinguished by certain features: automatism; sensitivity or sympathy, the effect of nerves, and also sometimes of electrical force fields; potential for publicizing what is private or hidden; and usefulness, often simultaneously, in the worlds of both occult and technological communication. Each of these constitutes a conceptual vein that runs through this book. Significantly, though, the figure I have just described appears in narrative less often as a static trope than as an image to be borrowed from, manipulated, and even written against. In fact, one of the reasons literature is so useful for studying the female medium is that it thus enriches our picture of her. Contemporary stories become a space for negotiation: these texts circulate but also complicate the paradigm I am proposing, subjecting the ideal to narrative scrutiny, revealing its contingencies or deficiencies, and effectively widening the social conversation about mediation. These narratives have a crucial role, then, in helping us to discern the mediating woman's impact on the culture and the ideological contests that accompanied her rise.

To point to the complexity of the big picture is also to recognize that the more optimistic images and ideas I address coexisted with other representations of female media. In electricians' writings, for instance, women were sometimes characterized as wasteful users of devices like the telegraph and,

---

48. A recent essay by Bette London is a notable exception. London points to the career of Louise Owen—an early twentieth-century private secretary who received séance messages from her deceased boss, a former owner of a British tabloid—to claim links between séance mediumship and secretarial work: a common "appeal to both self-assertion and self-effacement, to professional status and inspired amateurism, to public service and private ministration," coupled with gendered "sympathetic skills"; "the qualities that make for an ideal private secretary are precisely the same as make for an ideal medium—a sensitive typewriter, a 'machine' with intelligence and feeling." London concludes that through spiritualism, Owen sustained her role as intimate and discerning interpreter of her boss's will, representing him to others and enabling his continued outreach to the masses, in the process absorbing, like other mediums to renowned figures, some of his influence and celebrity for herself. See "Secretary to the Stars: Mediums and the Agency of Authorship," in *Literary Secretaries/Secretarial Culture,* ed. Leah Price and Pamela Thurschwell (Aldershot, UK: Ashgate, 2005), 92, 100, 101. My work adds to this astute discussion by proposing that the status of a woman like Louise Owen as a sensitive machine fundamentally derives from her function of communicating her boss—his meaning, his self—to others. That is, I want to emphasize that the "qualities that make for an ideal private secretary" or "ideal [spirit] medium" arise out of a structural communicative position. Investigating the situation from this angle furthers our understanding of both relayed communication (the goals for it; the anxieties around it) and femininity (the traits considered so essential to it that they became generalizable in the figural mediating woman; the implicit literary debates about the validity of those traits).

as operators, incompetent and garrulous.[49] To differing degrees, operators and typists might also be eroticized, and their non-domestic vocations cast them in the problematic mold of the New Woman.[50] Yet, I suggest, the development of a multiply networked culture presented challenges—to ideas of gender, dialogue, intimacy, privacy, and the relationship between body and self—too numerous and profound to be summarily resolved in terms of any one figuration. The mediating woman inspired and responded to a variety of priorities as well as preconceptions; and while the anxieties I have mentioned certainly existed, the figure of the sympathetic, non-interfering medium constituted a fantasy that competed with them. Again, the texts I investigate help us to see these moments of conflict and thus to develop a fuller portrait.

Another conceptual vein in the book has to do with literature itself as a relayed—mediated or channeled—communication. Occasionally plotted or implied acts of channeling, such as séance or mesmeric channeling, become metaphors for *textual* channeling—that is, most often, for the narrator's conveyance of the story from the author to the reader. Analyzing such textual relays allows for intriguing insights into the reception of text and narrative, especially when we focus on the position and tonal characteristics of the narrator-medium. We see for example how narrative mediation, like other instances of mediation explored in the book, can provide a locus for sympathy, constituting an affective bridge out to the reader (and not coincidentally this effect is allegorized through female characters). The narrator's station then becomes a potential instrument of rhetorical appeal—and, at the worst, manipulation—as it can be used by authors to cultivate desired emotions in the reading audience.

Each chapter in the book addresses fictional imaginings of some aspect of dialogue exchange or knowledge transmission and the communicative utility of the feminine channel. Chapter 1 elaborates the nervous sensitivity of the woman medium and its significance within communication networks. Henry James's *In the Cage* at once tacitly acknowledges and illustrates the

---

49. Carolyn Marvin, *When Old Technologies Were New: Thinking about Electric Communication in the Late Nineteenth Century* (New York: Oxford University Press, 1988), 22–31. See also Katherine Stubbs, "Telegraphy's Corporeal Fictions," in *New Media, 1740–1914*, ed. Lisa Gitelman and Geoffrey B. Pingree (Cambridge, MA: MIT Press, 2003), 97–99.

50. On suspicions about the sexual purity of the operator, see Marvin, *When Old Technologies Were New,* 26–29, and Stubbs, "Telegraphy's Corporeal Fictions," 98–99. Allen's *Type-Writer Girl* explores the typist's association with New Woman ideals, as does George Gissing's more grim *The Odd Women* (1893). But in a sense, the typist became a more palatable version of the New Woman: the discourse around her, by emphasizing her low wages and status as an object of male desire, ultimately reinforced her relativity to and reliance on men; see Christopher Keep, "The Cultural Work of the Type-Writer Girl," *Victorian Studies* 40 (1997): 401–26.

romance of the notion of mediumistic sympathy by rendering it in a more perverse form. Economically much less fortunate than the wealthy aristocrats she serves, the telegrapher-protagonist of the novella nonetheless thinks of herself as intrinsically sharing in their nobility. This thinking leads her, by turns, to fantasies of an ethereal rapport with her customers—a rapport James sketches through spiritualistic metaphors—and to a paradoxically selfish solicitousness about their affairs. The telegrapher's material conditions complicate her affective responses: sympathy takes on egoistic outlines because born of a desire to partake in lives from which she is excluded, a desire that motivates her infringement on the privacy of others' messages.

Chapter 2 considers such intrusions further. Frequent claims for the mediating woman's automatic or mechanical behavior intimate that this valuably compensated for a sensitivity that by itself could amount to a worrisome over-interest in others' transmissions. Women's supposedly ready mental nullity secured conversation from interlopers. Ironically, though, within certain horror fictions, it creates conditions of insecurity by allowing an inroad for incursion by Eastern foreigners adept at magically controlling unconscious states. Analyzing Bram Stoker's *Dracula* (1897), Grant Allen's and May Cotes's *Kalee's Shrine* (1886), and Marie Corelli's *The Soul of Lilith*, I claim the thematic significance of the mediating woman in late Victorian Gothic literature, as both a vulnerable site of intercultural conflict and a figure whose acute sympathy could be reshaped to effect appalling allegiances to the invading culture. At the same time, *Dracula* as well as Corelli's writings challenge the notion of the automatic woman itself as an inaccurate representation of women's intellectual aptitude.

The first two chapters address the medium's threat to privacy, while the next two address her functionality in desired acts of publicity or discovery. Automatism did not just render women ignorant of what they transmitted; it converted them into unresisting bodies and thus all the more efficient media in others' projects of diffusing (sending) or gathering (receiving) knowledge. My third chapter examines another instrument of Oriental cultural penetration, the automatized heroine of George Du Maurier's *Trilby* (1894), who becomes a broadcast or mass medium of the virtuoso Svengali's musical genius. Indeed, the novel anticipates the views of mass media and mechanical reproduction of twentieth-century writers like Walter Benjamin. Notably, too, the lucrative success of *Trilby*'s song turns on a sympathetic voice that similarly characterizes the novel's own mediating voice—its narrator—such that the heroine's performance offers a self-reflective program for *Trilby*'s own outstanding success, its triumph with a mass audience. Chapter 3 also underlines Du Maurier's depiction in all of his novels of the phonographic unconscious: an image of latent human memory or perception as a recording technology on the model of the phonograph or camera. In examining *Peter Ibbetson* (1891) and *The Martian* (1897), I am especially interested in how Du Maurier links the phonographic unconscious to matters of human evolution, our past as well as future development, representing it as a gateway to knowledge of our spiritual immortality. Like many writings characterizing the unconscious as a subliminal form of consciousness—a perspective still too often historically undervalued in the wake of psychoanalysis—these novels rebut Kittler's insinuation that turn-of-the-century technologies encouraged perceptions of the mind as self- or soul-less.

Chapter 4 looks at narratives that picture the automatized woman as an instrument of discovery, a "detective device." As a knowledge channel, this figure contributes vitally to literal detective stories' generic narrative trajectory from secrecy to revelation. In Arthur Conan Doyle's "A Case of Identity" (1891), the dim-witted typist is virtually interchangeable with the machine she works and which Sherlock Holmes uses to track down his antagonist; while in other stories, the stereotype of inattentive typing itself ironically enables women's own acts of detection by providing them a cover for their investigations. American psychologist Morton Prince offers up another kind of detective story in *The Dissociation of a Personality* (1905), a popular case study of what was later termed multiple personality disorder. In this condition lay yet another occasion for women, the usual recipients of the diagnosis, to mediate the communications of other selves. Prince's study ostensibly recounts the medical discovery of personalities unconscious to his pathologically nervous patient. Yet its contrived narrative elements and engaging literary and spiritualistic imagery, together with his hypnotic molding of his patient's behavior, mean that his claims of discovery constantly compete with antithetical signs of his creativity.

The final chapter considers the sporadic literary appearance of men who serve as occult relays. I account for this gender anomaly by emphasizing these characters' emasculation; Robert Browning's "Mr. Sludge, 'The Medium'" (1864), George Eliot's *The Lifted Veil* (1859), and Rudyard Kipling's "Wireless" (1902) all connect the role of psychic sensitive to personal or social impotence. My other objective, particularly in reading *The Lifted Veil,* is to further explore the communication channel as a figure for literary transmission. T. S. Eliot's "Tradition and the Individual Talent" (1919) depicts both male mediation, tinged with occultist imagery, and literature as a transmissive act. Eliot reconceives the male channel for the purposes of an implicitly masculine poetics; yet ultimately that re-gendering only highlights, by contrasting, the general persistence of feminine mediation well into the twentieth century.

This book illuminates ideas about women and communication that were significant enough to extend to different nations and different types of literature. There were of course some discrepancies between British and American technological and literary practice. The British Post Office assumed control of the telegraph and telephone after a few decades of each technology's invention, while in the United States the two remained in the hands of private industries. In the United States, telegraphy was employed mostly by businesses, whereas in Britain, where nationalization helped to make it more affordable, as much as two-thirds of messages were personal correspondence by the late 1800s, and British telegraph usage substantially exceeded American usage. Partly because the telegraph had become so entrenched in Britain, the competing technology of the telephone was adopted less enthusiastically there than in the United States, which saw relatively widespread telephone service in both businesses and homes by the early twentieth century.[51] In the realm of the occult,[52] Britain was the birthplace of the Society for Psychical Research, and one could argue that with the late Victorian British Gothic trend, magical events dominated British fictions more than American ones. Both factors underlie the greater attention to British writers on the paranormal in this study.

Despite some national variations of expression and degree, technological progress and interest in the occult were pronounced in both Britain and the United States, and the female medium occupied a notable place in transatlantic literature and society. Likewise, she makes a mark in all registers of fiction. To have limited analysis to only canonical or highbrow texts would

51. For a discussion of different patterns of development of telegraphy and telephony in America and Britain, see Starr, *Creation of the Media,* chap. 5 and 6.

52. As may be clear by now, my analysis of the "occult" does not encompass the turn-of-the-century movements sometimes designated by that term, such as Theosophy and neo-Hermeticism, which looked to Eastern or older Christian lore as a means to mystical and magical adeptness. The groups in question diverged from modern spiritualism (and mesmerism) in their attitude toward women, affording them places of leadership and intellectual agency, as in the case of Helena Petrovna Blavatsky, the co-founder and first president of the Theosophical Society; Annie Besant, her successor; and Anna Kingsford, the president of the London Theosophical Society and founder of the Hermetic Society. And in general, women were respected as fellow learners with men in the hierarchies on which these societies were built. Perhaps more crucially, these occultists had a fundamentally different relationship to spiritual knowledge than did modern spiritualists. whereas the latter emphasized the mediated route and communications with the dead, the former—even while sometimes intersecting with spiritualistic pursuits—emphasized the individual's study of esoteric teachings. In short, this version of occultism did not have at its core an idea of feminine instrumentality within communication relays and thus falls outside the scope of the communicative paradigm this book explores. For a thorough treatment of this occultism, see Alex Owen, *The Place of Enchantment: British Occultism and the Culture of the Modern* (Chicago: University of Chicago Press, 2004).

have meant ignoring her universality and the range of textual agendas she served. Moreover, it would have meant adhering to categories problematic in themselves in that they create artificial divisions that obscure similarities—for example, between Corelli's and James's works, both of which cast doubt on the verisimilitude of an image of femininity, and between *The Lifted Veil* and *Trilby,* both of which consider the reader's experience of narrative voice. Examining the woman as relay requires traversing paths that diverge in some places, intersect in others, producing an intricate network of language, culture, and belief.

❧ CHAPTER 1

Sympathy and the Spiriting
of Information *In the Cage*

If there was one note sounded most frequently
in nineteenth-century discussions of the electric telegraph, it was ebullience
at its promise of far-flung community. A popular history of the electric telegraph compares the marvel of connectivity of our own age. The "Victorian
Internet" was, as its contemporaries put it, a network of "human sympathies"
encircling the earth and making it "palpitat[e] with human thoughts and
emotions."[1] The language of the meeting of minds and hearts, which turned
up repeatedly at this time, implies something unexpected caught in this first
Web: its achievement was conveying not just information but something more
ineffable—actual human inwardness. This attitude may indeed help to explain why the telegraph often became spiritualized (and spirits became telegraphic) within the intertwining discourses of occultism and technological
speculation. Though the telegraph especially inspired claims of a dawning
age of unity, other Victorian-born communication machines could also seem

---

1. Quoted in Tom Standage, *The Victorian Internet: The Remarkable Story of the Telegraph and the
Nineteenth Century's On-line Pioneers* (New York: Walker, 1998), 163, 82. On the telegraphic news
press as a lauded instrument of sympathetic national unity following the shooting of an American
president, see Richard Menke, "Media in America, 1881: Garfield, Guiteau, Bell, Whitman," *Critical
Inquiry* 31 (2005): 638–64. On the joys as well as dangers of the intimacy seemingly created by technological as well as occult communications, see Pamela Thurschwell, *Literature, Technology and Magical
Thinking, 1880–1920* (Cambridge: Cambridge University Press, 2001).

to lend themselves to ethereal connections—as when, in a ghost story by Algernon Blackwood, the telephone permits the last desperate reach of feeling and personality of a drowned husband making a postmortem call to his wife.[2] Such phantasmal musings suggest a faith in communication technologies' ability to clutch and transfer the soul.

Yet these sanguine perspectives tell only part of the tale. Victorian communication technologies, along with movements like spiritualism, helped to inaugurate a modern notion of "communication" itself, with its strivings for a kind of dialogue wherein individuals transmit their interiority to each other. Yet this fantasy soon faltered, as people struggled with gaps between self and other and with means for negotiating them—with language, bodies, bureaucracies, and machines that facilitated mutual understanding but also, ironically, got in the way of it.[3] In other words, there was a problem with media, those intervening agents that paradoxically enabled and impeded communication. It is worth underscoring that many media were, or were operated by, human beings, such as séance mediums and telephonists. My aim is to spotlight the nineteenth-century gendering of human mediation, explicating it as a strategy for addressing the type of communicative concerns at issue here. If media potentially annoyed by their in-betweenness, *female* media mitigated that annoyance by seeming to offer special virtues that aided interpersonal contact. The feminization of mediation is, precisely, a response to that annoyance: it attempts to maximize the connecting powers of the apparatus while, conversely, minimizing its existence as a hindrance.

In what follows I address the first of these two benefits of feminization, examining notions of women's affectively sensitive bodies as these seemed to underlie conversational networks. This benefit had its limitations, though, as I begin to reveal in my reading of Henry James's *In the Cage* (1898). The trope of the mediating woman was one that literary works brought into focus but sometimes, as in the case of James, imagined only to amend or interrogate. In its coyly double-edged representation of mediated messaging, *In the Cage* attests to James's literary attraction to psychical research. But above all, as a realist piece, it emphasizes its telegrapher-heroine's painful awareness of her socioeconomic circumstances, which makes fulfilling the comforting vocational stereotype I describe an impossibility or, more exactly, necessitates that

---

2. Algernon Blackwood, "You *May* Telephone from Here," in *Ten Minute Stories,* 170–78 (1914; Freeport, NY: Books for Libraries, 1969).

3. John Durham Peters, *Speaking into the Air: A History of the Idea of Communication* (Chicago: University of Chicago Press, 1999). Thus, Peters's dazzling argument continues, communication as a goal ultimately flourished alongside fears of communication breakdown and solipsism, generating the obsession over failed connections that figures so prominently in Modernist writings.

she effectively refashion it for her own problematic ends. What is even more problematic, her actions clarify a pitfall of the stereotype itself for an ideal of relayed knowledge exchange. While some writers sought to legitimate the medium's work, James as well as others accentuated the dangers she brought to the vocation or the vocation brought to her. Making our way through this complex terrain demands beginning with the most foundational narratives, about women's feeling contributions to communication networks.

Imagining the Sympathetic Relay

> The humblest hello-girl along ten thousand miles of wire could teach gentleness, patience, modesty, manners, to the highest duchess in Arthur's land.
>
> —Mark Twain, *A Connecticut Yankee in King Arthur's Court*

In 1870, sometime journalist Justin McCarthy published a story in *Harper's New Monthly Magazine* reflecting on a technology that had proven a boon to journalism, the electric telegraph. Judging by this fiction, though, what captured McCarthy's imagination was not telegraphy's journalistic applications but rather the women so often employed to mediate it. "Along the Wires" tells the story of Annette Langley, a young woman working as an operator in "one of the great Atlantic cities."[4] Though unremarkable physically, she has other charms, most notably an ability to imagine what others feel, an intuitiveness that allows her to spin tales about them, to "throw herself into the lives and joys and sufferings of others, and thus put away her own petty vexations for the hour" (416). The plot turns on Annette's sympathetic imagination, especially as stimulated by her work: "Every one who came with a message to the office was compelled, quite unknown to himself, to tell her his story—or at least the story which his face, his expression, his voice, his message, and *her* fancy all combined to tell for him.... No doubt she guessed truly in many cases, for she was a quick, sharp, sympathetic girl" (416). This story-making about others' lives serves primarily to brighten and spice up her own; but there is also something sincere and touching about her sympathy, enough for one of her customers, a doctor named Childers, to take notice. Childers is surprised to find this trait in a woman, but we are

---

4. "Along the Wires," *Harper's New Monthly Magazine* 40 (1870): 416. All further page references to "Along the Wires" are parenthetical within the text. Thomas Jepsen pinpoints the story as McCarthy's in *My Sisters Telegraphic: Women in the Telegraph Office, 1846–1950* (Athens: Ohio University Press, 2000), 119.

given to understand that his surprise stems from his narrow conceptions of the emotions, the product of studying them "too much with the eyes of an intellectual anatomist" (417). Nonetheless, he rightly appreciates Annette's sensitivity, and the gist of the tale concerns his tests of it as he dispatches romantic-sounding telegraphs just to watch the play of emotions on her face while she processes them. "Good Heavens," he concludes, "what a sympathetic heart this poor girl has! And what tenderness and thoughtfulness one may see in her eyes!" (418).

Tender Annette Langley resembles a professional type that would come on the scene in the next decade, the female telephone operator. But this type took a few years to emerge, because when American switchboards were first getting underway around 1880, managers hired boys to operate them. That decision was soon regretted; the boys were rowdy as well as rude to callers, cursing at them and even threatening them with violence, so managers replaced them with young women. A similar pattern of employment trial and error occurred in the United Kingdom, such that on both sides of the Atlantic, all daytime telephone operating had been reserved for women by the late 1880s.[5] Apparently, what gave women the edge over the boys was not just their efficiency but also their kindly bent toward others. In 1938, one chronicler of the telephone industry summed up employment trends by observing, "the work of successful telephone operating demanded just that particular dexterity, patience and forbearance possessed by the average woman, in a degree superior to that of the opposite sex."[6]

Good switchboard conduct enjoined more than just "patience and forbearance"; in some instances, it involved operators in acts of extreme self-sacrifice. Another early history of the switchboard recalls Mrs. Mildred Lothrop, who in 1920 became the first recipient of AT&T's Noteworthy Public Service award for having braved floodwaters to inform local citizens of their danger. Faced with that same situation back in 1908, Mrs. Sarah J. Rooke had been less fortunate: after warning her subscribers, she herself was swept away and drowned. AT&T did not establish their public service award until 1920; yet, we learn, "the spirit of devotion for which [operators] were awarded was by no means new": "Human needs existed, then as now. Then, as now, human hearts and hands responded to them."[7]

---

5. On American employment, see R.T. Barrett, "The Changing Years as Seen from the Switchboard," *Bell Telephone Quarterly* 14 (1935): 52; on British employment, see F.G.C. Baldwin, *The History of the Telephone in the United Kingdom* (London: Chapman and Hall, 1938), 269.

6. Baldwin, *History of the Telephone,* 269.

7. Barrett, "Changing Years," 292. For more on the notions of women's submissiveness, patience, and altruism surrounding switchboard work, see Michèle Martin, who argues that the notion that

The operator's full heart and outreached hands embodied the premium quality she brought to her work: sensitivity or sympathy, gendered feminine by the early twentieth century, and marking out the operator as one of a larger class of mediating women.[8] The National Telephone Company in Britain viewed the operator's almost preternatural sensitivity to others as an imperative: a recruitment leaflet specified that applicants should have "insight into knowing what people mean to say when they cannot say what they mean to say."[9] Likewise, a feminine ethic of personal care was crucial to the image the Bell System sought to cultivate in its operators in the early decades of the telephone industry. In a policy that was simultaneously gender-, class-, and race-based, managers hired only white, genteel, and virtuous young women who would be attentive to callers' preferences and moods. As Venus Green explains, disappointing a caller's expectations required the operator's careful emotional response; when she had to report a line was busy, "it would be with a sympathetic tone to convey 'I am *sorry*, Mr. Smith, but I cannot give you what you want.'"[10] Social and moral attributes as a basis for personally attuned service, not simply managerial aims of minimizing wages or labor unrest, accounted for the feminization of the Bell switchboard.[11]

The gently patient "hello-girl" in Twain's *Connecticut Yankee* occupied the same ranks as telegrapher Annette Langley, as did eventually the woman at the typewriter. "A girl, such as I have in mind," affirmed one writer on the subject of typists in 1890, "has her eyes about her, she is full of sympathy, constantly on the alert for unpleasant things which she may avert or turn to good account. She anticipates the wishes of her employer and gratifies them almost before he has them."[12] Within a burgeoning turn-of-the-century

operating was a moral "labour of love" for women contributed to its feminization by seeming to frame them as easily disciplinable; *"Hello, Central?": Gender, Technology, and Culture in the Formation of Telephone Systems* (Montreal: McGill-Queen's University Press, 1991), 60. Also on feminine stereotypes in operating, see Kenneth Lipartito, "When Women Were Switches: Technology, Work, and Gender in the Telephone Industry, 1890–1920," *American Historical Review* 99 (1994): 1075–1111.

8. Peters likens the telephone operator to the spirit medium and to Melville's "Bartleby the Scrivener," concluding that all exhibit a "passive, neutral or feminine gender identity"; and he describes her, as someone who sends communications "across the chasm," as a representative of eros or desire for another (*Speaking into the Air*, 196). But the gender identity of the operator was far from neutral, as I underscore, along with the particularly Victorian-defined sympathy that seemed to underlie her work.

9. Quoted in Christopher Browne, *Getting the Message: The Story of the British Post Office* (Dover, NH: Alan Sutton, 1993), 128.

10. Venus Green, *Race on the Line: Gender, Labor, and Technology in the Bell System, 1880–1980* (Durham: Duke University Press, 2001), 68.

11. Ibid., 57–59.

12. Quoted in Carole Srole, "'A Blessing to Mankind, and Especially to Womankind': The Typewriter and the Feminization of Clerical Work, Boston, 1860–1920," in *Women, Work, and Technology:*

corporate ethos of community and subordination, women's fabled sympathy and good-heartedness could make them seem valuable workers in general. In the secretarial field specifically, female workers were sometimes described as more sensitively mannered and more governed by emotion than male ones. While some deprecated this higher affective investment as a workplace frailty, it seems significant, on the other hand, that what might be called an extreme of sympathetic identification soon became a staple of the job: popular and professional discourse regularly depicted the private secretary as someone who took on the boss's identity, becoming his "alter ego." Private secretaries were at the same time expected to craft their own work, which was, along with interpersonal abilities, what eventually raised them above the strictly typist and stenographic ranks. Thus secretaries faced paradoxical demands, being expected to demonstrate both worker self-erasure and worker independence.[13] Another way to put this paradox is to note that even as employers called for the skill of personal ingenuity, they also harbored an ideal of the secretary's mind as devoid of personal consciousness, because a perfect replica of the boss's. It is the latter kind of situation, incidentally, that makes feminine sympathy theoretically compatible with feminine automatism.[14]

Whether averting a disaster (a flood, a missed appointment) or intuiting her boss's needs, the operator or secretary worked best by working responsively and feelingly. This was a task well suited for women, as women had been described by that paragon of Victorian womanhood, Sarah Stickney Ellis. Ellis writes of the female creature,

---

Transformations, ed. Barbara Drygulski Wright et al. (Ann Arbor: University of Michigan Press, 1987), 95.

13. On secretaries in the turn-of-the-century corporate world, see Angel Kwolek-Folland, *Engendering Business: Men and Women in the Corporate Office, 1870–1930* (Baltimore: Johns Hopkins University Press, 1994), esp. 55–60. The cachet of secretarial work as opposed to mere typing and stenography was not immediate; in the last decades of the nineteenth century, typist-stenographers executed the same work later associated with secretaries, with a main difference being the size of the offices where they worked. The status hierarchy emerged in the early twentieth century as typing and stenographic jobs became more standardized and as women of families of lower social class began adopting them (Kwolek-Folland, *Engendering Business,* 65–66).

14. Intriguingly, in a 1909 essay, the mind of the private secretary is described as a phonograph: "As nearly as he can be described, he is a man who has lost his own personality and found his chief's in its stead. His brain is a plastic fac-simile of his chief's; indeed, like a piece of wax that has been molded to another form, it is so shaped as to think exactly as the chief thinks... he is, in other words, a sort of mental phonograph that never plays its own tune, that never originates but copies perfectly, that furnishes the chief with another extra brain" (quoted in Kwolek-Folland, *Engendering Business,* 60; her ellipsis). This image of the optimal private secretary—here designated by a masculine pronoun but increasingly female—strikingly resembles phonographic accounts of the unconscious mesmeric subject; see my chap. 3. The similarity attests to the conceptual links at this time between different forms of automatized (and feminized) communications.

She enters, with a perception as delicate as might be supposed to belong
to a ministering angel, into the peculiar feelings and tones of character
influencing those around her, applying the magical key of sympathy
to all they suffer or enjoy, to all they fear or hope, until she becomes
identified as it were with their very being, blends her own existence
with theirs, and makes her society essential to their highest earthly
enjoyment.[15]

As Ellis conceives it, sympathy has a specific direction; it draws a woman outward, until her own self "blends" with other selves. It is this move outward
that made women so appropriate for communication systems: their other-
directed presence was a linking force, creating bridges to users and, with herself as a nexus, between them. A principal component of the secretary's job,
like that of the operator's, was to make connections: she needed to become
her boss's alter ego so that she could accurately communicate him to others,
whether through typing or other means. At the least, positioning women
at crucial medial points facilitated smooth business relations. At the most, it
bettered the chances of realizing fantasies that communication technologies
could indeed generate a network of "human sympathies."

Such visions of breachless communication benefited by the presence of
wires and other connective devices like those found in electrical networks;
the rhetoric of sympathy surrounding the secretarial worker, which emerged
latest, was possibly an analogical elaboration of that already in place to describe operators in these networks. At the other end of the timeline was that
early and most spectacular species of Victorian communication medium,
the spirit medium. In Ellis's work, the depiction of women as "ministering
angels" whose sympathy pushes to the limit others' "earthly enjoyment" indicates the easy conceptual leap from the Victorian true woman to the spirit
medium. Ellis's understanding of sympathy as a "magical key" is also telling.
The concept of sympathy had long figured in discussions of magical or occult
intercommunications, beginning with the rise of mesmerism in the late eigh-
teenth century.[16] Women especially were said to enter readily into a sympathetic rapport with their mesmerists, and by the peak of modern spiritualism,
a similar aptitude was being used to explain their receptivity to the spirits'
will. Central to spiritualistic pursuits was the séance medium or "sensitive,"

---

15. Sarah Stickney Ellis, *The Women of England: Their Social Duties and Domestic Habits* (London:
Fisher, [1839]), 203.

16. Adam Crabtree, *From Mesmer to Freud: Magnetic Sleep and the Roots of Psychological Healing*
(New Haven: Yale University Press, 1993), 61–63, 73–74, 226–27.

a term implying important presumptions about the minds and bodies of the women who were mediumship's usual practitioners. The medium was a passive instrument, well attuned to the subtle cues, sometimes described as vibrations, by which the spirits expressed themselves. These perceptive responses were supposed to originate in fine nerves, which the Victorians believed characterized women's constitutions more than men's; therefore successful occult communication—spiritualistic as well as mesmeric/hypnotic and telepathic—resided in the idiosyncrasies of female neural biology.[17] In the absence of wires, women's impressionable nerves supplied the apparatus essential to here-to-hereafter communications.

McCarthy's "Along the Wires" implies the fluid associations in this period between the sympathetic spirit medium, with her nervous accouterments, and the sympathetic operator. According to Dr. Childers, it is sympathy that lies behind reported occult phenomena: "Not one man in twenty thousand can take into his sympathies what another man feels. . . . There is hardly any limit to the insight of a fine, sensitive, sympathetic, and at the same time scientific nature, which can at once feel with the feelings of others and see with the eyes of itself. I have no doubt all your sorceries, witchcrafts, second-sights, spiritualisms, mesmerism, and so on are to be explained in this way" (417). A skeptic, Childers nonetheless ironically approximates a chief tenet of modern spiritualists: that strange and wordless communications are explicable in material terms as the workings of sensitive constitutions. He thinks it exceptional for someone to enjoy this degree of sensitivity—or rather for a "man" to do so; significantly, his reference to the generic individual shifts once he imagines someone actually being that sensitive: "Some rarely endowed man *or woman* has the faculty of opening and using eyes and heart together, and dull people, who can not believe in any body seeing naturally what they themselves can not see, straightway invent supernaturalisms to explain what is simple nature unspoiled" (417; emphasis added). The shift in diction insinuates McCarthy's view—not Childers's yet—that sympathy is a feminine attribute, besides adumbrating the doctor's eye-opening acquaintance with the woman who relays his messages.

Annette has the talent for occulted perception that Childers has theorized, and it may well derive from her sensitive nerves. After she has missed work

---

    17. Roger Luckhurst discusses the nervously sensitive occult woman in *The Invention of Telepathy: 1870–1901* (Oxford: Oxford University Press, 2002), 215–19. See also Ann Braude, *Radical Spirits: Spiritualism and Women's Rights in Nineteenth-Century America* (Boston: Beacon, 1989), 83, and R. Laurence Moore, *In Search of White Crows: Spiritualism, Parapsychology, and American Culture* (New York: Oxford University Press, 1977), 106, 121.

for a number of days, Childers discovers her at home "prostrate with a severe nervous attack of a nature which he hardly understood." All he can make out is that she has "a highly nervous organization, and that 'something was on her mind.'" This "something" comes down, we might say, to a surfeit of sympathy: Annette has finally fallen in love with Dr. Childers and her emotions now overwhelm her. He, a tad behind as always, suspects the root of her nervous attack but not himself as its cause: "I tell you...she's in love....That's her secret—that's her ailment. She has an exceptionally sensitive and delicate organization—and she's in love with some fool or other" (420).

The story ends conventionally enough: Annette and Childers marry. Yet the blithe segue from sympathy to love should not distract us from what "Along the Wires" has to teach about popular perceptions of women media. Even Childers's comic over-intellectualizations give us insight into Victorian pseudo-scientific notions of telecommunication. Before realizing that he loves Annette, and hence puzzled by his intense interest in her affairs, Childers concludes that she must be emitting some kind of electromagnetic energy: "certain sympathetic organizations affect each other by the evolution of electric currents. He was not quite clear whether the brain, the heart, or the spinal marrow was to be regarded as the battery which set the currents in motion" (419). On one level, this explanation subtly compares what occurs in the telegraph office to spiritualistic communications. As believers asserted, those communications only seemed immaterial because they resulted from invisible electromagnetic currents extending from the medium's séance to the spirit realm. On another level, Childers tacitly likens Annette's "sympathetic organization"—rooted figuratively in her heart, literally in the nerves connecting her brain and spinal marrow—to the organization or network within which she mediates. Both types of organization function as electrified conduits of communication, such that Annette's nervous body becomes a miniature version or perhaps coextension of the telegraphic system.

Seeing her body in this way would have been reasonable at a time when people often spoke of telecommunication networks in terms of neural networks. For many, the nerves' swift conveyance of information appeared to be an apt model for the transmission of telegraphic messages; not only did nerves, like telegrams, communicate ideas and sensations quickly to remote points, they were now being said to do so by means of electrical signals.[18] Thus the nervous system became a preferred figure for the nineteenth-century

---

18. Laura Otis, *Networking: Communicating with Bodies and Machines in the Nineteenth Century* (Ann Arbor: University of Michigan Press, 2001).

electric telegraph.[19] Even after Marconi had done away with its wires, it still reminded people of nervous relays; in 1899, John Trowbridge remarked that thanks to the wireless telegraph, the "nerves of the whole world are, so to speak, being bound together, so that a touch in one country is transmitted instantly to a far-distant one."[20] Similar claims occasionally hovered around the telephone. Bell's invention had transformed the industrialized city into a living web of communication pathways: "From the great ganglion familiarly hailed as 'Central' radiate the myriad nerves along which speed the impulses directing the world's industry."[21]

These neural metaphors affirm a desire to see the electrified world in utopian terms: as one body, united by common intents and sensations. The trope of the mediating woman underwrote this desire; installing neurally sensitive women at the "great ganglion" of the "Central" switchboard or telegraph office mounted a defense against the potential discomforts of an increasingly technological landscape. The altruistic intermediary sustained the hope that, far from dividing and isolating (as we sometimes still fear they do today), modern technologies would forge connections, no matter the separations created by distance or situation, including—in spiritualism—death itself. Thus, too, an apparent obstruction to person-to-person contact, the medium's necessary interposition, was made less of one, if not indeed a positive factor toward human intimacy.

Genteel Affinities:
Etherealizing the Message in James

In McCarthy's story, Annette's sympathy rewards itself, winning her a husband and pulling her out of the life in which it has been put to so much use. Not that that life has been entirely without diversions, for sympathy is what powers her active imagination:

> If some of the utterly commonplace people who went in with their absolutely uninteresting and prosaic messages could only have known what striking central figures of romantic story she made out of them they would have been a good deal surprised, and many of them, probably,

19. James W. Carey, "Technology and Ideology: The Case of the Telegraph," *Prospects: An Annual Journal of American Cultural Studies* 8 (1983): 314.

20. John Trowbridge, "Wireless Telegraphy," *Appleton's Popular Science Monthly* 56 (1899): 72.

21. "Behind the Scenes at 'Central,'" *Booklovers Magazine* 2 (1903): 391.

would be very angry. No doubt she guessed truly in many cases, for she was a quick, sharp, sympathetic girl; and many sad stories are hinted clearly enough even in the briefest telegram. (416)

Eventually the narrator draws back to the crux of the tale, Annette's sympathy, but not before the first part of the passage has stuck in one's head. Even that quick moment raises certain questions. Why would Annette's sympathy surprise her customers? Why would it anger them? McCarthy does not pursue these questions; he is mostly committed to painting Annette in a rosy light. Yet the logical response here is that there is something intrusive about her interest in others' lives, especially given that her sympathy is always accompanied by a knack for keen observation: "Annette looked at [Dr. Childers]. She always looked at every body" (417). The romances she constructs about her customers are, says the narrator, just "harmless fictions" (416). But to what degree are they really harmless, or even fictions, when she has based them on accurate inferences from telegrams? In short, where does sympathy end and eavesdropping begin?

If McCarthy is not inclined to take this up, another writer dealing with similar material is. Like "Along the Wires," Henry James's novella *In the Cage* (1898) concerns a young "sympathetic" telegrapher who passes the time by creating stories about her patrons and who begins to focus on the telegrams of one attractive gentleman in particular.[22] But ultimately these superficial resemblances only bring out McCarthy's and James's diverging investments in the same story, ones to some degree based on differences of genre and tone. James rewrites McCarthy's slight romantic tale in a realist mode, significantly building on what are only bare intimations in McCarthy's narrative of the medium as, in Childers's terms, a "good, true girl" (418). *In the Cage* accentuates that that narrative veils the probable inadequacies, even ugliness, of the telegrapher's existence. It is not that James's heroine does not try to be a good, true girl, but rather that her efforts in that line continually run up against the material and moral shortcomings in her way of life. What results is a rendering of female mediation in which the medium's sympathy is fully exposed as a gilding of her troublesome curiosity about others' life stories, a curiosity fueled by her deep-seated repudiation of her class status.

Much of the emotional difficulty faced by *In the Cage*'s heroine has to do with the embarrassed social position of the turn-of-the-century female telegrapher. Although social gradations ultimately arose between, say,

---

22. Jepsen also notes the similarity between the two works (*My Sisters Telegraphic*, 138).

private secretaries and mere typist-stenographers, generally speaking, the new technological forms of communication mediation—telephone operating, telegraph operating, and typing—reserved for women some measure of white-collar prestige, because they occurred indoors, were cleaner than factory labor, and required some education and a learned skill.[23] Some employers clearly sought to amplify this prestige by fashioning a workplace as simulacrum of the bourgeois woman's domestic environment. Bell Telephone, for instance, styled itself as a familial unit and exerted a strict parental discipline over its switchboard operators. These young women worked under the supervision of "matrons" who advised them in matters of hygiene and dress and oversaw finely decorated retiring rooms, where operators could retreat during breaks to read a magazine or play the piano.[24] As we can conclude, motherly supervisors and opportunities for leisure safeguarded the gentility of women workers by camouflaging a public workspace as a private bastion of middle-class life. Equally useful for creating this illusion was the insularity of the switchboard: lunchrooms first came into being so that women would have the option of remaining within the exchange, avoiding the ignominy of encountering men on the street.[25] The Chicago Telephone Company put the matter baldly when, in a pamphlet about its operating school that included photographs of rooftop gardens and of "recreation rooms" furnished with "comfortable chairs, couches, reading tables," and a circulating library, it boasted "that such environment attracts and keeps the better type" and that telephone operating as a whole was "free from direct and sometimes unpleasant association with the public... since here the employees are shielded from direct personal contact."[26]

Conversely, for women in telegraphy, the status of their work could be compromised by a condition particular to many of their jobs: their frequent

---

23. On class considerations, see, e.g., Meta Zimmeck, "Jobs for the Girls: The Expansion of Clerical Work for Women, 1850–1914," in *Unequal Opportunities: Women's Employment in England 1800–1918,* ed. Angela V. John (Oxford: Blackwell, 1986), 155, 158; Gregory Anderson, ed., *The White-Blouse Revolution: Female Office Workers since 1870* (Manchester: Manchester University Press, 1988), 10, 42–43; Stephen Norwood, *Labor's Flaming Youth: Telephone Operators and Worker Militancy, 1878–1923* (Urbana: University of Illinois Press, 1990), 28, 45–48; and Edwin Gabler, *The American Telegrapher: A Social History, 1860–1900* (New Brunswick, NJ: Rutgers University Press, 1988), 57–58, 85–91.

24. Norwood reads these cozy workplaces as Bell Telephone's attempt to stave off unionization and collective protests by increasingly Taylorized employees (*Labor's Flaming Youth,* 48–52). If this was Bell's motivation, it is still worth noting how much it catered to class-based standards of femininity.

25. Brenda Maddox, "Women and the Switchboard," *The Social Impact of the Telephone,* ed. Ithiel de Sola Pool (Cambridge, MA: MIT Press, 1977), 268.

26. Chicago Telephone Company, *Operators' School: First Lessons in Telephone Operating* (Chicago: Chicago Telephone Company, 1910), 15, 11.

interaction with strangers—and strange men specifically—at the public telegraph office. As the telephone companies' secluded workplaces remind us,
"the better type" of young woman of this era was marked not just by certain
economic practices but also by premarital sexual integrity. In questions of
social respectability, then, the usually unwed female telegrapher faced a moral
criterion largely irrelevant to her male colleagues. The fact of her publicized
body made it easier to view her along a continuum with the prostitute, effectively debasing her to a lower class position than her profession in and of
itself should have earned her.[27]

"Here, indeed," declared an 1883 contributor to the *Pall Mall Gazette* on
the subject of switchboard operating, "is an occupation to which no 'heavy
father' could object; and the result is that a higher class of young women
can be obtained for the secluded career of a telephonist as compared with
that which follows the more barmaid-like occupation of a telegraph clerk."[28]
As if in direct response to this association with the wanton mingling of the
tavern worker, the narrator of *In the Cage* states of the telegrapher-heroine
that "she believed in herself... if there was a thing in the world no one could
charge her with it was being the kind of low barmaid person who rinsed
tumblers and bandied slang."[29] This statement, which appears in a discussion
of the heroine's attraction to Captain Everard, sheds valuable light on her
self-estimation: though realizing "the picture of servitude and promiscuity"
she presents at Cocker's, "so boxed up with her young men" (194), she retains a sense of honorable difference from the working-class woman and that
figure's putative sexual easiness.

Several critics have observed the telegrapher's potential alignment with the
prostitute, especially in her dealings with Captain Everard.[30] My argument
starts by underscoring her deliberate rejection of this identification. Rather
than countenance selling her body, she attempts to shift in her own mind the

27. On the possible confusion between the public urban woman worker and the prostitute, see
Sally Ledger, *The New Woman: Fiction and Feminism at the Fin de Siècle* (Manchester: Manchester University Press, 1997), chap. 6. On female telegraphers' extraordinary publicity and stories about their
sexual openness, see Katherine Stubbs, "Telegraphy's Corporeal Fictions," in *New Media, 1740–1915,*
ed. Lisa Gitelman and Geoffrey B. Pingree (Cambridge, MA: MIT Press, 2003), 98. Gabler records
other factors militating against American female operators' claims to gentility, among them working-
class backgrounds and poor living conditions (*American Telegrapher,* 128–29).

28. Quoted in Peter Young, *Person to Person: The International Impact of the Telephone* (Cambridge:
Granta, 1991), 27.

29. Henry James, *In the Cage,* in *In the Cage and Other Tales,* ed. Morton Dauwen Zabel (Garden
City, NY: Doubleday, 1958), 204. All further page references to *In the Cage* are parenthetical within
the text.

30. See for example Eric Savoy, "'In the Cage' and the Queer Effects of Gay History," *Novel* 28
(1995): 284–307.

dimensions of her relations to the aristocracy from a worldly to non-worldly, or other-worldly, plane of connection. Like McCarthy's, James's tale subtly gestures toward analogies between telegraphic and occult communications. *In the Cage* appropriates but also meaningfully reorients spiritualistic and other magical imagery to delineate the tormented class consciousness of the telegrapher and her longings to rise above her degraded social position.

Appreciating these ideas demands recognizing how and how much James's works were influenced by late nineteenth-century occultism.[31] James became familiar with psychical research through, among other avenues, his brother William, who was pivotal in establishing the American branch of the Society for Psychical Research. This exposure, Martha Banta suggests, deepened Henry's authorial interests in the occult, in ways that sometimes had less to do with identifiable occurrences than with understated manipulations of occult themes as a means of exploring human relations and perceptions. He was drawn, for example, to depictions of feminine sympathy, orienting his stories around the "divulgence of concealed information by psychically sensitive women able to see and thus to know and to reveal all."[32] With *In the Cage,* I propose, James twists this favorite motif to ironic ends, portraying a woman who only imagines herself a psychic sensitive and in a position to know and reveal all.

In the first pages of *In the Cage* when Lady Bradeen drops in to Cocker's grocery, the narrator states of her that the "apparition was very young, but certainly married" (181). The word *apparition* crops up again later in the context of the floral arranger Mrs. Jordan's vaunting discussion of her clientele, in which the telegrapher feigns only partial interest: "There was something in our young lady that could still stay her from asking for a personal

---

31. It is a testament to *In the Cage*'s weird suggestiveness that several critics have written on its occult elements. However, none has offered a sustained analysis of the story's magical and especially spiritualistic themes. Within an argument about choice and ethics, Janet Gabler-Hover classifies the novella among James's ghost stories on the basis of its use of *apparitions* to describe affluent characters. But she is more interested in reading the telegrapher as a mental vampire, someone who unconsciously preys on others to broaden her own scope of existence, than on exploring the story's ghostliness; "The Ethics of Determinism in Henry James's 'In the Cage,'" *Henry James Review* 13 (1992): 265–68. For T. J. Lustig in *Henry James and the Ghostly* (Cambridge: Cambridge University Press, 1994), the telegrapher's message mediation and subjective expansion form part of the story's uncanny themes (191–93). In a provocative juxtaposition, Pamela Thurschwell suggests that the interclass intimacy the telegrapher enjoys doubles that experienced by James's typist Theodora Bosanquet, who was interested in psychical research and purportedly channeled her employer after his death (*Literature, Technology,* chap. 4). But while noting the telegrapher's implicitly telepathic sympathy, Thurschwell otherwise leaves strangely out of account the novella's occult imagery.

32. Martha Banta, *Henry James and the Occult: The Great Extension* (Bloomington: Indiana University Press, 1972), 159. Also on the female sensitive in Henry James's writings, see Luckhurst, *Invention of Telepathy,* 234–51.

description of these apparitions; that showed too starved a state" (193). On both these occasions, *apparition* denotes people of the highest class of society, the aristocrats, inhabitants of "homes of luxury" who can afford to lavish money on copious flowers and copious telegrams (192). As aristocrats, Mrs. Jordan's customers, along with Lady Bradeen, embody a relation to history precisely encapsulated in the idiom *apparition*. Like ghosts, they symbolize a past that asserts itself in the present: they are visible reminders of centuries-old principles of class and community inherited by late Victorian London. Thus the telegrapher, when she first serves Lady Bradeen, sees in her face an entire legacy—"her birth, her father and mother, her cousins and all her ancestors" (180). If this woman is a phantasmal emblem of persons now departed, *In the Cage*'s modern setting registers both the departure and what is progressively replacing it: the capitalist economy of the city, as figured in the grocer Mr. Mudge. Lady Bradeen's face presents to the telegrapher a wraith of eminence dwindling away amid the countless throngs and transactions of the fin de siècle metropolis.

But Lady Bradeen ghosts more than just the history of British culture. She also hearkens back in her gentility to the telegrapher's own particular family history. Although James leaves uncertain how prestigious a class the telegrapher once belonged to, he does make clear her descent from some higher stratum and her perception of herself as a lady *manquée*. The narrator tells of the era in the heroine's life when she, her mother, and her sister found themselves, "conscious and incredulous ladies, suddenly bereft, betrayed, overwhelmed," then "slipped faster and faster down the steep slope at the bottom of which she alone had rebounded" (176). Together with Mrs. Jordan, an acquaintance "handed down from their early twilight of gentility and also the victim of reverses" (177), the telegrapher has suffered disgraces only appeased by a sense of ingrained nobility: "It had been a questionable help, at that time, to ladies submerged, floundering, panting, swimming for their lives, that they *were* ladies; but such an advantage could come up again in proportion as others vanished, and it had grown very great by the time it was the only ghost of one they possessed" (191). The story's "apparitions" personify for the telegrapher just this sort of ghostly remembrance of the higher opinion she once commanded.

Neither she nor Mrs. Jordan entertains lasting illusions of their formal gentility. Nevertheless, a principal subplot of the story involves the two women's jockeying for a greater sense of proximity to the upper class, a proximity that they can sometimes suppose exceeds their impersonal service positions and passes over into a more rarefied realm of intimate knowledge in which their patrons appreciate, despite the women's distressed circumstances, their enduring dignity. This dream is fundamentally a dream of transcending material differences, and fittingly James conveys it through images of spiritual or psychical connections.

Early in the story, there is an anticipatory playfulness in the narrator's comparison of the telegrapher's communication mediation to that of a sporadically summoned oracle: "there were long stretches in which inspiration, divination and interest quite dropped" (178). Her interest is not truly piqued until she encounters Lady Bradeen and Captain Everard. After a brief exchange of business with them, she imagines that they still hover around her, like displaced spirits, throughout her tedious workday:

> They remained all day; their presence continued and abode with her, was in everything she did till nightfall, in the thousands of other words she counted, she transmitted, in all the stamps she detached and the letters she weighed and the change she gave, equally unconscious and unerring in each of these particulars, and not, as the run on the little office thickened with the afternoon hours, looking up at a single ugly face in the long sequence, nor really hearing the stupid questions that she patiently and perfectly answered. (183)

The telegrapher's preoccupation with her favorite aristocrats, coupled with her "unconscious" yet "perfect" completion of her work, suggests her ability to revert to automatism, of the sort practiced by automatic writers at séances, in which an intervening "presence" takes priority over the conscious self. Like a medium's, her mind has been given over to her "apparitions," and gradually, we learn, her "divinations work faster and stretch farther." Her station in the cage allows her what the narrator describes as a "queer extension of her experience" or "double life": a feeling of closeness to a formally remote (class) stratum (186). When Lady Bradeen visits Cocker's again, her features seem to glow with thoughts, probably of Everard, that give the telegrapher "the sharpest impression she had yet received of the uplifted, the unattainable plains of heaven, and yet at the same time caused her to thrill with a sense of the high company she did somehow keep" (211).

Somewhat incongruously, this mystical extension is enabled by a crude voyeurism and by her ha'penny novels, whose sentimental plots intimate the possibility of her own social rise through rescue by Everard.[33] Yet crucially,

---

33. Nicola Nixon analyzes the delusiveness of the telegrapher's romantic novels, arguing that she ultimately experiences a realist awakening; "The Reading Gaol of Henry James's *In the Cage*," *ELH* 66 (1999): 179–201. For other arguments on the telegrapher's relation to realism, see Patricia

her outlook is not purely idealistic: her fantasy is complicated throughout the story by her painful understanding of her actual life conditions. Even her emotions for Everard are never wholly idyllic. James keeps the contradictions she embodies—naiveté and awareness, a desire to save and a tendency to condemn—in constant tension, and yet these are unified by the idiom of occult sensitivity.

The telegrapher comes to believe she is channeling Everard's thoughts immediately, without the burden of verbalization, and this is part of her romance: the reverie of psychical communication seems based in her hopes that he implicitly recognizes her inner nobility. Importantly, it evolves from the meeting with Captain Everard in the Park that she has angled for as a "miraculous" occasion for the display of her honor: "All our humble friend's native distinction, her refinement of personal grain, of heredity, of pride, took refuge in this small throbbing spot" (210). But that very optimism bespeaks her more troubling mindfulness of what Everard might otherwise think. It is remarkable how often the unspoken relays that take place during this meeting confirm the fact that she is not a woman who would prostitute herself. The encounter begins with an exchange that immediately cuts short what she fears to be his perception of her streetwalking: "Are you taking a walk?" "Ah I don't take walks at night! I'm going home after my work" (217). The exchange is an oblique one, and in fact, the narrator implies it may only be silently "smiled out" (217). The obliquity and silence are key, because the telegrapher prefers their not having to utter "anything vulgarly articulate" on the subject of her honor: "She had an intense desire he should know the type she really conformed to without her doing anything so low as to tell him" (217, 219). What seems in the first place "vulgar" about explanatory speech is the initial horrid misapprehension of her character that it would presume. The idea that Everard can intuit that she is not a prostitute magically elevates their relationship above the most conspicuous material determinant of her low-class status, her provocatively exposed body, which, coupled with the other material determinant, her poverty, would seem to give the impression of availability for the right price.

It is not that prostitution does not arise during their meeting as a tacit possibility but rather that that possibility is, again tacitly, denied. After she announces that she has not yet eaten, she seems overjoyed that he does not offer

---

Walton, *The Disruption of the Feminine in Henry James* (Toronto: University of Toronto Press, 1992), 91–100, and Richard Menke, *Telegraphic Realism: Victorian Fiction and Other Information* (Stanford: Stanford University Press, 2008), chap. 6, the latter of which incisively reflects on the telegrapher's mediating position.

her a meal: "she at once felt sure she had made the great difference plain. He looked at her with the kindest eyes and still without saying what she had known he wouldn't. She had known he wouldn't say 'Then sup with *me!*' but the proof of it made her feel as if she had feasted" (219). The invitation to sup is a loaded element in the story. While normally it would provide an occasion for intimacy, because Everard already has a lover (Lady Bradeen), it stands for the telegrapher as a sign of merely casual and short-term physical gratification. In other words, as she earlier realizes, men of his class might take women of her class as women who do not really "count as infidelity" (218). His not asking her to sup seems, then, an indication that she has made her "great difference" from the latter group "plain."

Days later, back at her cage, she imagines that the Park meeting has established an understanding so profound, so spiritual, that the two need no longer bother to "clumsily to manoeuvre to converse... the intense implications of questions and answers and change, had become in the light of the personal fact, of their having had their moment, a possibility comparatively poor. It was as if they had met for all time—it exerted on their being in presence again an influence so prodigious" (236). In a pinch, this fantasy of affinity can even dispel the humiliation of evidence to the contrary: evidence of his indifference or, worse, lack of esteem for her character. Thus she declines to interpret his "putting down redundant money" as an attempt to buy her and instead supposes that, reading into his thoughts, she gleans an entire range of more flattering meanings: "He wanted to pay her because there was nothing to pay her for. He wanted to offer her things he knew she wouldn't take. He wanted to show her how much he respected her by giving her the supreme chance to show *him* she was respectable. Over the dryest transactions, at any rate, their eyes had out these questions" (242). The supposition of psychically channeled communications is vital for rescuing the telegrapher's pride, and yet this is also clearly an ambivalent narrative moment insofar as it reveals some doubt on her part, a point I will return to later.

It is crucial that her contact with him rise above not only money (a purchase price) but also verbalization, because as her profession dramatizes, language itself is caught up in worldly systems of privilege and exchange. Money and words in this novella are intertwining modes of currency, both implicated in existing social relations, as Dale M. Bauer and Andrew Lakritz remark.[34] Moreover, as James's heroine has come to know them, words rudely

---

34. Dale M. Bauer and Andrew Lakritz, "Language, Class, and Sexuality in Henry James's 'In the Cage,'" *New Orleans Review* 14, no. 3 (1987): 65. These critics argue that the telegrapher imagines herself in a virginal, salvational role in Everard's life above both sexual intrigues and class limitations.

confront her with a material reality that conflicts with her most hopeful esti-
mations of her own worth, the reality of sudden penury, which now requires
her to "count words as numberless as the sands of the sea" (174). Before the
Park meeting, the telegrapher fancies she is communicating with Everard in
an ethereal way that even when voiced defies language by defying its strict
significations: "no form of intercourse so transcendent and distilled had ever
been established on earth. Everything, so far as they chose to consider it so,
might mean almost anything" (204–5).

If, on the one hand, the telegrapher dreams she enjoys with Everard a rapport unsullied by articulation, on the other hand, her communications with
her tradesman fiancé are laughably verbose. The "daily deadly flourishy letter
from Mr. Mudge" brings the telegrapher crashing back to the level of material difference, where the paths of human lives—for instance, her own desperate engagement—often submit to the direction of economic facts (178).
Mudge's crass rootedness in the worldly realm manifests itself even in his
diction, wherein words are weighed down by a certain heaviness—by those
"present, too present, *h*'s" (175). Further, his longwinded utterances are rendered in accounting metaphors. About their vacation plans, Mudge "flooded
their talk with wild waves of calculation," handling "the whole business . . . as
a Syndicate handles a Chinese or other Loan" (215). On the vacation the
telegrapher finds herself more tolerant of this endless chatter—including his
"perpetual counting" of the people on the Bournemouth pier—because this
gives her the chance to mull furtively over her own "secret conversations":
"This separate commerce was with herself; and if they both practised a great
thrift she had quite mastered that of merely spending words enough to keep
him imperturbably and continuously going" (230–31). In the story's telegraphic logic, which equates language with money, Mudge's obsession with
words is congruent with his obsession with the capitalist order, and being
with him induces the telegrapher to sink earthward and brood in mundane
terms of "commerce" and "spending."

Mudge's diction divulges his (botched) effort to set himself apart from the
h-dropping laborers that constitute his family background, and by this token
we gather that he is as innately allied to the working class as his fiancée is
to some upper one. Now the two occupy comparable social strata, yet the
telegrapher chafes that he can be "so smugly unconscious of the immensity
of her difference" from him (198). Were she romantically attached to Everard, she thinks, matters would stand quite otherwise, with such "a relation

---

Also on the telegrapher's desire for a relation with Everard outside the financial/linguistic nexus, see
Jennifer Wicke, "Henry James's Second Wave," *The Henry James Review* 10 (1989): 146–51.

supplying that affinity with her nature that Mr. Mudge, deluded creature, would never supply" (208). Indeed, the absence of affinity between her and Mudge comes out in the sheer volume of words that pass between them, as this reveals their lack of silent empathy, on which her relation with Everard subsists. Mudge has his own mediatory connection to the aristocracy, yet this mocks the telegrapher's fantasy of infiltrating the genteel psyche. Whereas she imagines an almost spiritual channeling of ideas and emotions, for him it is wealth that gets channeled from upper to lower levels of society. As he believes, the riches of the nobility nourish the general economy and indirectly become the boon of average working individuals. There is only promise and profit in his career as middleman: "the exuberance of the aristocracy was the advantage of trade, and everything was knit together in a richness of pattern that it was good to follow with one's finger-tips" (202).

By contrast with Mudge, Mrs. Jordan shares in the telegrapher's fitful dream of an ethereal link to the aristocracy. Like the telegrapher, the floral arranger has suffered setbacks that have reduced her to wage-earning, but this state of affairs has not kept her from assuming her abiding fellow feeling with genteel society. By her account, her exceptional origins have given her a rare insight into the aesthetic preferences of the aristocracy, such that in her business it is gentlemen who are "her greatest admirers; gentlemen from the City in especial" (253). Mrs. Jordan is sure that her good breeding differentiates her from other workers: "The regular dealers in [floral] decorations were all very well; but there was a peculiar magic in the play of taste of a lady who had only to remember, through whatever intervening dusk, all her own little tables, little bowls and little jars and little other arrangements, and the wonderful thing she had made of the garden of the vicarage" (190). Importantly, this "peculiar magic"—Mrs. Jordan's extraordinary facility with flowers—resembles a well-known Victorian type of magic, one associated with spirit mediums. While *In the Cage* frames the telegrapher's work as a species of automatic writing, Mrs. Jordan's profession recalls a more visually sensational high Victorian style of mediumship. Many séance leaders acquired fame for producing palpable objects out of thin air (or more precisely, as believers asserted, out of the domain of spirits). These so-called *apports* might consist of any number of articles the spirits desired to manifest, but the most common variety seems to have been flowers.[35] Mrs. Jordan's floral productions recall

---

35. Thus for instance in Robert Browning's "Mr. Sludge, 'The Medium'" (1864), the sitters expect the medium will eventually "Make doubt absurd, give figures we might see, / Flowers we might touch." Robert Browning, *The Poems,* ed. John Pettigrew, vol. 1, 821–60 (London: Penguin Books, 1981), ll. 411–12.

this feat, particularly when one considers her custom of ministering her services to the invisible. As she admits to the telegrapher, while she is conjuring up her arrangements, her aristocratic patrons are "nearly always out" (193). Hence they exert a presence for her still more ghostly than that of her friend's aristocrats, who at least regularly materialize at Cocker's grocery.

At first the telegrapher cannot hear of Mrs. Jordan's business without thinking of the dead, because "her one idea about flowers was that people had them at funerals" (194–95). Mrs. Jordan's clients have yet to depart this world; nonetheless there is something weirdly mediumistic in her dealings with them, as for example when she describes her decorative preparations for their dinner parties as a medium might describe her function at a séance circle: "They simply *give* me the table—all the rest, all the other effects, come afterwards" (192). By exhibiting her work as a kind of mediumship, the narrative registers the full depth in her opinion of her intuition of aristocratic psychology. Her floral arrangements, she says, often come down to "ineffable simplicities," matters of taste that resist expression or that men like Lord Rye do not bother to detail but rather only "[throw] off—just [blow] off like cigarette-puffs— such sketches of." Here, she proudly suggests, she is in effect channeling the wishes of her gentlemen; they silently will what is to appear, and she infallibly actualizes that will. In the absence of Lord Rye's explicit wishes, Mrs. Jordan finds herself dependent on her "imagination" or, like so many channels before her, her "sympathy" (197).

Centripetal Sympathy

The persistent paradox beneath this supposed spiritual tie to the aristocracy is that Mrs. Jordan remains aware of her fixity in present material conditions, as does the telegrapher. Indeed, the pathos in James's telegrapher lies in his depiction of both sides of her "double life." *In the Cage* oscillates between what the telegrapher envisions to be her access to the genteel life and what she knows is her practical difference from it. Occasionally her remembrance of that difference evinces itself in bursts of pure antagonism: "What twisted the knife in her vitals was the way the profligate rich scattered about them, in extravagant chatter over their extravagant pleasures and sins, an amount of money that would have held the stricken household of her frightened childhood, her poor pinched mother and tormented father and lost brother and starved sister, together for a lifetime" (187).

The trope of mediumship proves revealingly agile in this regard, capable of expressing both poles of the telegrapher's ambivalence. Predictably the telegrapher has "wonderful nerves," and she sometimes considers lending her insight to her patrons, imagining this aid as a "hazard of personal sympathy" (177, 188).[36] On the one hand, this sympathy extends the telegrapher's fantasy, in that it amounts to an ability to enter readily into the psyches of her aristocratic customers, the "cream of the 'Court Guide'" (175). On the other hand, it stems from her thoughts of her social inferiority—and here it begins to look very little like sympathy, as we conventionally perceive it, at all.

> There were impulses of various kinds, alternately soft and severe, to which she was constitutionally accessible and .which were determined by the smallest accidents.... She had thus a play of refinement and subtlety greater, she flattered herself, than any of which she could be made the subject; and though most people were too stupid to be conscious of this it brought her endless small consolations and revenges. She recognised quite as much those of her sex whom she would have liked to help, to warn, to rescue, to see more of; and that alternative as well operated exactly through the hazard of personal sympathy, her vision for silver threads and moonbeams and her gift for keeping the clues and finding her way in the tangle. The moonbeams and silver threads presented at moments all the vision of what poor *she* might have made of happiness. (188)

What is striking about this passage is how dramatically it turns sympathy inside out. Ostensibly, as for instance Ellis describes it, a feeling directed outward toward others, sympathy becomes instead a means of satisfying the telegrapher's own needs, especially those born of her economic disadvantages. For one thing, it gratifies her need for self-respect: it is something about which she can "flatter herself," probably because it is due to her "refinement," which might allude as much to her inherent nobility as to her sensitive nerves. Further, her sensitivity encompasses both concern and its opposite: though at times it leads her to want "to help, to warn, to rescue," at other times it paves the way for "endless small consolations and revenges," with the difference dependent on how events move her. Yet both alternatives, it would seem, function "exactly through the hazard of personal sympathy"—with that "hazard" already laying the stress on sympathy's benefits and costs for herself.

---

36. For Otis, James's language here associates bodily and emotional interconnections with the telegraph system's, as if the telegrapher were a neural component within the network's figurative brain (*Networking*, 169–71).

There is a persistent reflexivity in the telegrapher's sympathy. Seeing her genteel patrons, she is really picturing herself; they enable her a "vision of what poor *she* would have made of happiness," one that recalls Audrey Jaffe's claim that the Victorian sympathizer mentally exchanges the other with a spectacle, a similarly circumstanced self-representation.[37] But the telegrapher's experience pushes sympathy ironically close to a jealous antipathy. In fact, her "wonderful nerves" prime her for "sudden flickers of antipathy and sympathy" both (177); the "impulses" to which she is "constitutionally accessible" are "alternately soft and severe" (188). Then again, as her vision suggests, antipathy may only be the flipside of sympathy, and together they refer back to her own desires.[38]

Clearly James is doing something unexpected with mediumistic sensitivity. Though preserving it as an index of the vocation, he expands upon it: the telegrapher may be attuned to her customers, but that attunement does not always manifest itself straightforwardly as kindliness. There is even a hint of this versatility in "Along the Wires," as McCarthy's narrator briefly mentions one emotion accompanying, but also a far cry from, Annette's sympathy: "many ... had her profound sympathy, or admiration, or pity, or hatred who never deserved any such sentiment on her part; and many had it who well deserved it, and never knew any thing about her feeling toward them" (416). Maybe Annette, an orphan too poor to afford a doctor when she falls ill, derives her "hatred" from the same source James's telegrapher does. But McCarthy's is essentially the tale of a good-hearted girl who gets a good-hearted ending (the well-off object of her affections treats her for free, then marries her). Contrarily, *In the Cage* rests on the idea that the worker who feels intensely may—especially given the social gulf between herself and whom she serves—feel intensely and negatively. "I hate them," the telegrapher exclaims one day to Mrs. Jordan. "They're selfish brutes. . . . They bore me to death." At Mrs. Jordan's competitive response—"Ah that's because you've no sympathy!"—the telegrapher only laughs dryly, "retorting that nobody could have any who had to count all day all the words in the dictionary" (196–97).

---

37. Audrey Jaffe, *Scenes of Sympathy: Identity and Representation in Victorian Fiction* (Ithaca: Cornell University Press, 2000). Jaffe describes speculative sympathy as motivated by the fluidity of socioeconomic identity under capitalism—an anxious awareness that one could become the degraded other. In James's story, of course, it is the telegrapher herself who occupies the lowly position, such that the other becomes an object of coveted, instead of dreaded, identification.

38. Jaffe comes to comparable conclusions about the relationship between sympathy and its apparent antithesis, *ressentiment*, noting that both imaginatively dissolve social boundaries (chap. 3).

Even when she does sympathize with the upper class, as we have seen, her motive (and probably Mrs. Jordan's also) is what sympathy does or can do for her. "She would have given anything to have been able to allude to one of [Everard's] friends by name,... to one of his difficulties by the solution. She would have given almost as much for just the right chance... to show him in some sharp sweet way that she had perfectly penetrated the greatest of these last and now lived with it in a kind of heroism of sympathy" (205). The narrator radically alters the image of the mediating woman, from a locus of centrifugal altruism to one of centripetal egoism: it is Everard's *recognition* of her sympathy that most matters to her; to "show" him her concern is to turn the focus back in her direction and demonstrate to him what a hero she really is. Sympathy provides a source of self-aggrandizement and a means to ingratiate herself with Everard and his class.

Or, more ominously, to indebt them to her. During their meeting in the Park, the telegrapher magnanimously tells Everard, "I'd do anything for you. I'd do anything for you" (223). But the rest of the conversation reveals that her "particular interest" in his telegraphic affairs has motivated a vigilance that is positively worrisome (221).

> "What you *do* is rather strong!" the girl promptly returned.
>
> "What *I* do?"
>
> "Your extravagance, your selfishness, your immorality, your crimes," she pursued without heeding his expression.
>
> "I *say!*" her companion showed the queerest stare.
>
> "I like them, as I tell you—I revel in them. But we needn't go into that," she quietly went on; "for all I get out of it is the harmless pleasure of knowing. I know, I know, I know!"—she breathed it ever so gently.
>
> "Yes; that's what has been between us," he answered much more simply. (226–27)

Like the narrator of "Along the Wires," James's heroine dismisses knowledge gained through telegraphy as "harmless" and couches it in a notion of sympathy. Yet Everard recognizes that that knowledge has generated something potentially damning between them—an opportunity for blackmail.[39]

---

39. In *George Eliot and Blackmail* (Cambridge, MA: Harvard University Press, 1985), Alexander Welsh suggests that James's novella showcases the possibility of eavesdropping within sophisticated communication systems like the penny post and especially the electric telegraph (55–58). On the story's relevance to the then-recent Oscar Wilde trials and an exposé of informant telegraph boys-cum-prostitutes, see Savoy, "'In the Cage' and the Queer Effects," and also Hugh Stevens, "Queer Henry *In the Cage,*" in *The Cambridge Companion to Henry James,* ed. Jonathan Freedman, 120–38 (Cambridge: Cambridge University Press, 1998). Also on blackmail in the novella, see Thurschwell, *Literature,*

Still, and importantly, that the telegrapher would in the end really seek to
profit monetarily by what she has learned through Everard's exchanges remains doubtful. Although she briefly imagines trumpeting her knowledge
to him—"Come therefore; buy me!"—her imagination ends at that point,
"the point of an unreadiness to name, when it came to that, the purchasing
medium. It wouldn't certainly be anything so gross as money, and the matter
accordingly remained rather vague, all the more that *she* was not a bad girl"
(207). The telegrapher's refusal to blackmail is worth dwelling on because it
again illuminates her social self-image, along with her disdainful view of the
fallen world in which she is forced to operate. To extort money from Everard
would not be to establish the kind of relationship she desires but instead to
corroborate what those in his set are already disposed to think of her: that
she is simply low-class, a "bad girl." In addition, the potential "purchasing
medium" for her knowledge forms part of a material nexus that she shuns
here and elsewhere as "gross": blackmail would entail succumbing to the
base reality of her economic want. It would also mean resorting to a system
in which money and language work in tandem—blackmail borrows the
linguistic substance of the telegram, as Eric Savoy says[40]—the very kind of
system that employs her but denies her its enjoyments. So, while whatever
information she may have originates in language (in telegrams), she will not
reverse the operation by telling or threatening to tell what she has learned.

The undercutting of the prospect of blackmail does not fully do away
with the sinister aspect of her eavesdropping. What it does do is confirm
the degree to which the telegrapher's sense of worth and power lies beyond language and telling, in the realm of thought and knowledge. As
she sees it, her knowledge of Everard's actions provides her a covert—and
dematerialized—"possession":

> There were those she liked and those she hated, her feeling for the lat-
> ter of which grew to a positive possession, an instinct of observation
> and detection.... She had at moments, in private, a triumphant vicious
> feeling of mastery and ease, a sense of carrying their silly guilty secrets
> in her pocket, her small retentive brain, and thereby knowing so much
> more about them than they suspected or would care to think. (187)

With Everard, too, she basks in the "extraordinary possession of the elements
of his life that memory and attention had at last given her" (204). Knowledge

---

Technology, 97–98, as well as Andrew J. Moody, "'The Harmless Pleasure of Knowing': Privacy in the
Telegraph Office and Henry James's 'In the Cage,'" *The Henry James Review* 16 (1995): 53–65.
    40. Savoy, "'In the Cage' and the Queer Effects," 294.

through eavesdropping—possessing—alleviates the anguish of her poverty by seeming to permit her an illicit grasp on genteel society, and thus it is continuous with her sympathy, likewise indulging her yearning to be close to the genteel mind.

Notably, the telegrapher's claims of possession are oddly in tension with the actual nature of what she claims to possess. For N. Katherine Hayles, *In the Cage*'s telegraph is part of an incipient economy of information, which differs from an economy of conservation, wherein things are necessarily had by some and kept from others. As file-sharing demonstrates in modern times, information—unlike matter and energy—is reproducible and shareable, subject only to the limitations of accurate transmission and access. Having accessed Everard's secrets, the heroine tries to help him and to carve out a relationship "beyond and above the regime of scarcity."[41] I suggest that in its potential lack of territoriality—in the sense of both possessiveness and material groundedness—this telegraphic information economy lends itself to the heroine's quasi-spiritual hope of communion with the aristocrats. And yet, that she continues to speak of information as something that can be possessed indicates her vague sense, however tamped down for now, that one's power always comes down to the things one can call one's own. It is only a matter of time and circumstance until she regains a full appreciation of the vise-like grip of the economy of haves and have-nots and of her own function within it.[42]

Possessing the Telegrapher

Beyond the reach of speech, in her own mind, the telegrapher preserves a sense of her personal value. However, in her psychical fantasy, as we have seen, she entertains the hope of successfully transferring that sense to another person, Captain Everard. For this to be realized, thought would need to be externalized between, instead of locatable within, single persons—which in fact, according to Sharon Cameron, describes how thinking works in James's fiction. Thought existing between characters, she observes in analyzing one novel, can even take the "form of communication that looks curiously like

---

41. N. Katherine Hayles, "Escape and Constraint: Three Fictions Dream of Moving from Energy to Information," in *From Energy to Information: Representation in Science and Technology, Art, and Literature,* ed. Bruce Clarke and Linda Dalrymple Henderson (Stanford: Stanford University Press, 2002), 239.

42. For her part, Hayles finds the intransigence of the economy of scarcity in the telegrapher's refusal to share information with Mrs. Jordan at the end of the story.

mind reading."[43] Yet this communication begins to look less curious once we factor in James's authorial attraction to the paranormal and consider how even an apparently realist work like *In the Cage* might incorporate paranormal elements to portray characters' experience of their world. For Cameron, because thinking extends outside the individual mind, it can exert, or at least attempt to exert, a coercive force on other people; trans-subjective thought is potentially "asymmetrical" insofar as one character seeks to determine meaning for another.[44] Indeed, this is just what *In the Cage*'s telegrapher is trying to do when she imagines interchanging thoughts with Everard: the implicit purpose of the interchange is not really to strike up a harmony with him so much as to thrust upon him a recognition of her respectability. In effect, her fantasy works by dictating what she thinks she amounts to, what she means to herself, to his mind. But of course this fantasy upends her proper vocational role, which is to be dictated *to*—to be simply the conduit of others' thoughts and feelings. The reversal indicates the larger problem she represents in the story, which is that she is never sufficiently absent as a medium, instead obtruding intellectually on Everard and Lady Bradeen's communications.

However, what the telegrapher comes to realize is her inability to impose her interpretation of herself on others in any stable way. If the silent manipulation of meaning for another person turns Jamesian thinking into a political field,[45] the telegrapher's chimera starts to collapse once she perceives that this field cannot remain distinct from the politics figured through class and economic structures. With her increasing acceptance of her place in a culture where a person's value is largely established in and by a consumer economy, the asymmetry she imagines obtains psychically outside the linguistic/monetary order becomes inverted to reflect the asymmetry that obtains within it.

On the one hand, her unspoken exchanges with Everard prolong her dream that he thinks her worthy enough to give up a lady like Lady Bradeen for her if he only could. But on the other hand, these exchanges reveal the telegrapher's growing suspicion that a man of Everard's rank could only ever estimate her character and interest in him at a much lower rate. This suspicion has nagged at her all along, but it begins to become inescapable in the near-final scenes at Cocker's grocery, thoroughly adulterating her fantasy. For Everard's leering gaze begins to project just that meaning on her that she has resisted, conveying that if he desires to fulfill her longing in any way,

---

43. Sharon Cameron, *Thinking in Henry James* (Chicago: University of Chicago Press, 1989), 96.

44. Ibid., 109.

45. Ibid., 110.

it would only be in a casual, financially negotiable relationship. Recalling their Park meeting, she begins to wonder how he has interpreted her offer to help him:

> Mixed with her dread and with her reflexion was the idea that, if he wanted her so much as he seemed to show, it might be after all simply to do for him that "anything" she had promised, the "everything" she had thought it so fine to bring out to Mr. Mudge. He might want her to help him, might have some particular appeal; though indeed his manner didn't denote that—denoted on the contrary an embarrass-ment, an indecision, something of a desire not so much to be helped as to be treated rather more nicely than she had treated him the other time. Yes, he considered quite probably that he had help rather to offer than to ask for. (239)

The very inarticulateness of her "anything" has kept it open for his interpre-tive privilege as an aristocrat of choosing what she could possibly mean in his life. His looks seem to posit, if somewhat awkwardly, a trade in her favor: if she will only treat him especially "nicely," it is he who will have "help" to offer her in return, perhaps monetary help. Thus he replaces her tacit meanings with his own: "He had taken it from her in the Park that night that she wanted him not to propose to her to sup; but he had put away the lesson by this time—he practically proposed supper every time he looked at her" (241).

These unspoken communications are confirmed in that moment when he seems to be passing her some kind of surplus compensation: "It was either the frenzy of her imagination or the disorder of his baffled passion that gave her once or twice the vision of his putting down redundant money—sovereigns not concerned with the little payments he was perpetually making—so that she might give him some sign of helping him to slip them over to her" (242). Though it is possible to read the extra sovereigns as Everard's payoff for blackmail,[46] it is likely that she at least sees them as an advance pay-ment for prostitution. Whichever the case, in fact, the payment pegs her as a low-class woman with something immoral to sell, exactly the identity she sought to transcend. Is Everard really trying to buy her, or is this just what she envisions? The narrator's "either/or" commentary leaves the question open. If he is not, the telegrapher's anxious vision may be triggered by the

---

46. Moody, "'Harmless Pleasure,'" 63; Savoy, "'In the Cage' and the Queer Effects," 302.

escalating aggressiveness of her erotic fantasy. Just previous to these scenes at her cage, we see her returning to his neighborhood regularly in hopes of meeting him again, and even contemplating asking the doorman for entry to his residence (237). What she suddenly begins to read in his looks and gestures is precisely the most logical way a second encounter on these grounds would strike Everard: as an offering of herself, an impoverished telegrapher, for a type of infidelity that does not count.[47] Her change of mind is decisive: "She had passed his door every night for weeks, but nothing would have induced her to pass it now" (241). All that said, resolving what is happening with the sovereigns matters less than her perception of what is happening: that perception indicates an essential shift in her understanding—"the rush of a consciousness quite altered" (241)—of what she could really signify to a man like Everard, a man with far more leisure and money (or rather social connections to money) than she. Under such conditions, her dream that he might ultimately prefer her to a noblewoman like Lady Bradeen persists, yet becomes increasingly insecure: "how could she . . . know where a poor girl in the P.O. might really stand?" (243).

The telegrapher's awareness of her lack of control over meaning coincides with the dawning sense of her true ignorance of the aristocrats' lives. When Everard comes into Cocker's desperately hoping to retrieve a sent telegram, she can only wonder what "could be bad enough to account for the degree of his terror. There were twists and turns, there were places where the screw drew blood, that she couldn't guess" (250). The intertextual touch of the famously ambiguous *The Turn of the Screw*, published in the same year as *In the Cage*, reminds us that in both stories James uses the motif of the female medium's sensitivity to explore socioeconomic frustration, as well as the limits of interpretation.[48]

For all her confidence about knowing the secrets of the aristocrats, the telegrapher will never be nearer to them than she is now, as the caretaker of

---

47. We cannot assume, as Stuart Hutchinson does, that what the telegrapher fears is simply sex; "James's *In the Cage*: A New Interpretation," *Studies in Short Fiction* 19 (1982): 23–24. In fact, there are indications that the relationship she desires with Everard, though rarefied, does not necessarily exclude sex. The rarefied nature depends on transcendence not of the body in and of itself but rather of money, including the bodily encounters money can buy (and which her public vocation seems to advertise). During one of their conversations, the telegrapher fleetingly imagines herself and Everard "on a satin sofa in a boudoir"—as erotically involved, but in a high-class setting untainted by financial need—and does not balk when he reaches for her hand (223).

48. Moreover, as John Carlos Rowe points out, to the degree that she remains only a pawn in possibly nefarious aristocratic doings, the telegrapher resembles the governess in having a false sense of autonomy from the hegemonic order that employs her. *The Other Henry James* (Durham: Duke University Press, 1998), 162.

their correspondence—the position essentially of a glorified onlooker, and with far too limited a vantage point to ever bear out her boast of "possession." This is the epiphany *In the Cage* works toward, but it is also indicated throughout the text in the discreet metaphor of the telegrapher's mediumship and more specifically in the spectral mutability of the term *possession*. In effect, it is not she who possesses the aristocrats but rather they who, as "apparitions," have always had the benefit of possessing her. In less ghostly terms, this possession amounts to a statement of her position in the capitalist order: her financial straits have obliged her to sell her services to the British Post Office, which in turn has found the readiest buyers for those services in the aristocratic community. The telegrapher only adds to their options for selfish "squanderings and graspings" (187); to them, she represents just another object for their frivolous purchase and enjoyment.

Throughout the narrator connotes the telegrapher's possessed status through light rhetorical touches. Depriving her of a name allows James to reinforce her subordinate social status twice over: she is just an anonymous face for the people she serves, and her anonymity also requires the narrator to refer to her by a nondescript usage of the possessive case: "our heroine," "our young friend," "our young lady," and so forth. This usage increases dramatically in the final scene, when Mrs. Jordan has "possessed herself" of the heroine to announce her engagement to the butler Mr. Drake (255). Mrs. Jordan's possession seems a stand-in for the aristocrats', for it is through her that the telegrapher now learns of Lord Bradeen's death, Everard's financial troubles, and his pending marriage to Lady Bradeen: all news to "our heroine" and thus an indication of her real ignorance of high-end society. These disclosures steer her on to a "vivid reflexion of her own dreams and delusions and her own return to reality" (261).

Her dreams and delusions of a mystical link to the aristocracy give way to the gravity of material reality, as she accepts that in the world's opinion, she is virtually as marketable and possessible as that figure beneath her contempt for barmaids, the prostitute, with whom indeed Everard may have conflated her. It is this prospect that has bred her fear of him and resolved her that, "should it come to that kind of tension, she must fly on the spot to Chalk Farm," the neighborhood of Mr. Mudge (239). Her decision at story's end to do just that, to hasten her wedding, signifies her complete acknowledgment of the dangerous shallowness of her actual relations to the aristocracy. Contemporary documents reveal that wifehood and women's operating all but mutually excluded each other in the late Victorian era;[49] so one can

---

49. Charles H. Garland, "Women as Telegraphists," *Economic Journal: The Journal of the British Economic Association* 11 (1901): 259–60; Jepsen, *My Sisters Telegraphic*, 72.

assume that with her marriage, the telegrapher will resign her work not just at Cocker's but in telegraphy altogether. Rather than be taken for the aristocrats' possession, perhaps even a commodity for their sexual pleasure, she commits herself to a far more honored scenario—to being the conjugal possession of Mr. Mudge.[50] This is an outcome Mudge himself has always been certain of, regardless of his fiancée's slights, as his manner conveys: "Mr. Mudge presently overtook her and drew her arm into his own with a quiet force that expressed the serenity of possession" (235). With her husband, clearly, the telegrapher will not have evaded her status as possession. But at least she will have gained in the transaction financial security, as well as be spared the stigma endured by publicly employed young women. Marriage will do for the telegrapher what Bell Telephone tried to do for its switchboard operators—confine her body and protect her from the social misconceptions engendered by her sexual allure.

Most of the narrative's action has seen its heroine wedged between two extremes: the shelter of complete privacy (the domestic life offered in some indefinite future by Mr. Mudge) and the peril of complete publicity (the promiscuous and commercialized life offered in the ominous example of the prostitute).[51] The telegrapher's position between these two extremes is neatly figured by the titular cage, a structure intended to enclose but also, by its latticework, to make her readily accessible to the public. When she thinks herself sought after by Everard, she decides this cage has "become her safety" (241); nonetheless, it leaves her open to his mute entreaties. Interestingly, these moral risks to herself redouble other risks to the people she serves. As the physical node through which their dialogues travel, her vulnerable body becomes an apt synecdoche for the vulnerability of their messages; both represent potentially treacherous bridgings of privacy and publicity.

James is not the only author to exploit this doubleness. At one point in Anthony Trollope's "The Telegraph Girl" (1877), the protagonist Lucy Graham, who works in the telegraph department on an upper floor at the General Post Office, is sought out at work by her friend Abraham Hall. When Mr. Hall tries to enter the department to see Lucy, he learns from the doorkeeper "that the young ladies were not allowed to receive visitors during office hours."[52]

---

50. Here I agree with Bauer and Lakrtiz's description of the telegrapher's body as "sexually exchangeable" for economic comfort and their suggestion that her marriage constitutes the most socially legitimate mode of exchange ("Language, Class," 64). It is surprising in this context that they do not focus at any real length on the story's undercurrent of prostitution.

51. For Rowe, the telegrapher crosses the private-public divide in bringing motherly qualities to her telegraphic position (Other Henry James, 174–76).

52. Anthony Trollope, "The Telegraph Girl," in The Complete Short Stories, ed. Betty Jane Slemp Breyer, vol. 4 (Fort Worth: Texas Christian University Press, 1982), 97.

The doorkeeper's response invests the scene with erotic interest by describing Lucy's workplace in terms redolent of a place in which a romantic tête-à-tête would occur. Like a parlor, the telegraph department houses "young ladies" who can only "receive" male visitors during specified hours. In a sense, the scene replicates the logic of the switchboard by having the operator seem to occupy a space safeguarded from male attentions—largely by eliding the fact, mentioned earlier in the story, that the eight hundred women in the telegraph department work alongside some boys and young men.[53] The elision creates factitious protective boundaries (somewhat like the bars of a cage) between Lucy within the telegraph office and the possibility of sexual overtures coming from outside of it, as represented by Mr. Hall.

But the prospect of Lucy's issuing out from the department to chat with Mr. Hall poses more than the obvious dangers of publicity. "Now it is a rule," the narrator tells us,

> that the staff of the department who are engaged in sending and receiving messages, the privacy of which may be of vital importance, should be kept during the hours of work as free as possible from communication with the public. It is not that either the girls or the young men would be prone to tell the words which they had been the means of passing on to their destination, but that it might be worth the while of some sinner to offer great temptation, and that the power of offering it should be lessened as much as possible. Therefore, when Abraham Hall pressed his request the doorkeeper told him that it was quite impossible.[54]

The narrator first summarizes the rule that indicates an official anxiety that operators are likely to "tell the words" they transmit, then, disingenuously, denies that likelihood. If the denial succeeds in exculpating the operator, it does so only partially, by shifting attention away from the teller to the "sinner" who would tempt her to tell. Though the narrator imagines both female and male operators revealing privacies, the scene itself frames a woman, Lucy, as the potential teller—and Abraham Hall as her potential tempter. The erotic trappings of his visit, coupled with Lucy's exaggerated seclusion with other young ladies

---

53. Trollope notes this unsavory circumstance in his essay "The Young Women at the London Telegraph Office," even as his persona, touring the office, ironically submits the women to a sexualizing gaze: "But I was very anxious to know whether they—flirted, for there are young men in the same room. I thought that had I been a young man there I might have been tempted"; *Good Words* 18 (1877): 380.

54. Trollope, "Telegraph Girl," 97.

inside the office, suggest that his request that she leave it exposes her to another "great temptation" besides the one explicitly articulated by the narrator.

Even Mr. Hall, actually an upright fellow, recognizes the illicit sexual menace some might perceive him to embody. When Lucy, at last going out to meet him, acts hesitant about accompanying him any farther, he seems to interpret her hesitation as an unwillingness to be seen with a man on the streets: "Are you ashamed to walk with me?"[55] His interpretation, while mistaken (she is only ashamed of her ratty clothes), is logical given Lucy's earlier protest that he should stop sending a coworker of hers money because his presents give the impression that he expects some satisfaction in return. Lucy's earlier reflection that Sophy's "prettiness ha[s] its dangers and temptations" when it comes to her interactions with Mr. Hall implicates him in multiple temptations by the time he stands awaiting Lucy outside the telegraph office.[56] As far as Mr. Hall feels himself to be from the type of man who would take advantage of a young woman's poverty, Trollope's tale, like James's, nonetheless acknowledges the likelihood that others will confuse the telegraph girl—to the exact extent that she makes herself publicly available to male interest—with the most notorious working girl of all.

"The Telegraph Girl," also like *In the Cage,* overlays the sin of sexual publicity with another type, the sin of publicizing others' messages. When Lucy exits the telegraph office, she sets in motion two dangers simultaneously—as if her body were inseparable from the information she transmits. In fact, I am arguing, this is precisely the correspondence these narratives imply; and these twin themes, separately or together, may have been essential to portraits of female telegraphers. Fictions by male operators of the period sometimes picture the female telegrapher as jeopardizing the privacy of transmissions, at other times as subject to physical compromise—vulnerable to attack or licentiously exposed.[57] Stories like James's and Trollope's, wherein informational and bodily availability happen concurrently, give an especially vivid sense of how thoroughly women's sensitive bodies seemed to be involved in knowledge networks and transfers.

But James's narrative is as interesting for how it reshapes prototypical images of women media as for how it conforms to them. *In the Cage*'s heroine may feel a sympathy widely reputed for these media, but her sympathy is revealed as an oddly selfish selflessness, when not transforming altogether

---

55. Ibid., 98.
56. Ibid., 87.
57. Stubbs, "Telegraphy's Corporeal Fictions," 98–99. In Stubbs's view, such depictions were male operators' efforts to protect their jobs by discrediting lower-paid female operators.

into hostility. By dramatizing these feelings and the economic realities behind them, the novella implies how shaded a picture of the woman telegrapher must be to truly account for her affective relations with her much more privileged clientele. The dream of psychical attunement of James's heroine never fully eclipses her feelings of class abjection or her recognition of how her patrons must see her. Hence her emotions in her interactions with Captain Everard run a confused gamut: longing, resentment, vanity, shame, hope, fear.

The Perils of Sensitivity

In bending the operator's romanticized traits closer to reality, *In the Cage* shows that the narratives favoring the feminization of mediation in the long run generated an involved and conflicted conversation about that phenomenon. For instance, if women's sensitivity seemed a boon for telegraph offices and other technological "ganglions," it could also seem a bane. By many accounts, this sensitivity basically amounted to *over*sensitivity, or a tendency to "nerve strain," with significant consequences for the kind of work women could perform. In an article on female telegraphers, British economist Charles H. Garland referred to several sources attesting to the "natural disabilities of women," including a French expert who stated that the "natural nervous organisation of women is unable to adapt itself" to periods of heavy work. This quickness to strain and fatigue meant women were only suited for the lighter wires—those conveying personal and local messages rather than market and race reports—which required less skill and earned their workers less money.[58] The practice Garland describes of restricting women to second-class wires was common in telegraphy.[59] His comments indicate that suppositions of women's neural sensitivity helped to legitimate that practice, setting aside the more remunerative jobs for men and placing a ceiling on women's own rank and pay. In discussions of telephone operating, similarly, nerve strain as a concept worked to devalue women's paid employment and, by the same move, to underscore domesticity as a priority. In a 1909 analysis of working women, Elizabeth Beardsley Butler anxiously discussed an often cited report on Canadian telephone operators, quoting one doctor's pronouncement to the Canadian labor commission that after the operators "have gone on for four or five years and served the company, and they get married or for other

---

58. Garland, "Women as Telegraphists," 257–58.
59. Jepsen, *My Sisters Telegraphic*, 25.

purposes leave, then they turn out badly in their future domestic relations. They break down nervously and have nervous children and it is a loss to the community." For Butler, this report offered a warning to the American telephone industry to amend switchboard working conditions so as to avoid producing similar results by repeating "the same laboratory experiments with the nerve cells of its young girls."[60]

Plainly claims of neural incapacity were being used to reinforce desired economic and cultural scenarios, which, after all, had for some time typified the response to women seeking all manner of vocations outside the home. Nineteenth-century doctors theorized close links between the workings of women's nervous systems and their reproductive organs; they also envisioned the human body's nervous energy as a finite supply, on the model of the galvanic battery's. (Recall as an early version of this idea the electromagnetic organization of Annette Langley's nervously sympathetic body.) In this context it was easy to decry women's higher education or public employment as a drain on their nerve force with frightful consequences for their childbearing and child-raising capabilities.[61] But the situation was also more complicated than this in the specific case of communications jobs, for the operators themselves reported nervous problems: this was not simply a disorder imposed upon them by authoritative medical voices. Just what travails of mediation were these women expressing through this ambiguous diagnosis? According to an open letter on female telegraphers by the Women's Trade Union League of Illinois, "nervous strain often causes a paralysis of the right hand."[62] Compare this testimony with Butler's assertion about the switchboard that

---

60. Elizabeth Beardsley Butler, *Women and the Trades* (New York: Russell Sage Foundation, 1909), 289–90, 292.

61. Janet Oppenheim discusses the interlinking of the nervous and reproductive systems in doctors' cautions against women's higher education; *"Shattered Nerves": Doctors, Patients, and Depression in Victorian England* (New York: Oxford University Press, 1991), 187–201. John S. Haller, Jr., and Robin M. Haller recount similar concerns about women's involvement in business and other public pursuits; *The Physician and Sexuality in Victorian America* (Urbana: University of Illinois Press, 1974), 25–34; on the British scene, compare Jane Wood, *Passion and Pathology in Victorian Fiction* (Oxford: Oxford University Press, 2001), 163–85. On ideas of nerve force and neurasthenia, see Oppenheim, *"Shattered Nerves,"* chap. 3; Haller and Haller, *Physician and Sexuality,* chap. 1; and Tim Armstrong, *Modernism, Technology, and the Body: A Cultural Study* (Cambridge: Cambridge University Press, 1998), 15–17.

62. "The Lot of the Woman Telegrapher: An Open Letter" (Chicago: The Women's Trade Union League of Illinois, [1907?]), Compare the assertion by a British Post Office committee member that the difficulties of reading from the telegraphic sounder, "coupled with the closeness of the circuits to one another, and the consequently cramped position of the writer, results in increased strain on the nervous system"; Miss Mabel Hope, "Evidence on Behalf of the Female Telegraphists in the Central Telegraph Office, Counter Clerks and Telegraphists in the Metropolitan Districts, Returners at Mount Pleasant, and Telephonists at G.P.O. South" (London: Co-operative Printing Society, 1906), 6.

"managers and operators as a rule agree...that [women] get 'glass arm,' a nervous inability to work, more frequently" than men.[63] Thomas Jepsen, a historian of women's telegraphy, glosses "glass arm" as what we now call carpal tunnel syndrome.[64] In that condition, a nerve in the wrist is painfully compressed due to repetitive manual movements; so mentions of "glass arm" invite us to take operators' complaints of nerve strain pretty literally. Of course carpal tunnel syndrome remains an occupational hazard for word processors today, especially professional typists, and as such offers a physical sign of these women's descent from earlier female media.

But the unionists' complaint extends further, with telegraphers' nerve strain producing not only manual paralysis but also "a complete nervous collapse, and occasionally it produces insanity." Evidently nerve strain is not fully reducible to carpal tunnel syndrome; operators also resorted to the diagnosis to convey their jobs' psychological pressures. Butler bases part of her analysis of nerve strain on one switchboard operator's statement that "you nearly go crazy with the number of calls, and the supervisor at your back and the subscribers often so mean."[65] This last effect warrants a closer look: at least part of what made the telephone operator crazy (and the telegrapher insane?)[66] was having to deal with obnoxious customers who, in a telling reversal, tried the very qualities of patient endurance for which women media were known.

Even doctors agreed that this aspect of the job contributed to nerve strain. A 1911 *Lancet* article recounted the conclusions of a medical committee that had investigated telephone employees at the British Post Office: "a sound nervous system is essential, because the telephonist is constantly using three special senses—namely, speech, hearing, and sight" and "because the telephonist...deals directly with the public, 'whose knowledge of the method of working,' says the report, 'is limited and whose methods, manner, and temper are always diverse and sometimes unpleasant. She is constantly smoothing out difficulties, and is often the subject of abuse and reproach, whilst it is necessary for her to be businesslike, tactful, and courteous under all conditions.'" The medical report, the *Lancet* article continues, cannot propose adequate remedies for this "factor in the induction of ill-health in the telephone

---

63. Butler, *Women and the Trades,* 293.

64. Jepsen, *My Sisters Telegraphic,* 36.

65. Butler, *Women and the Trades,* 291.

66. The open letter on "The Lot of the Woman Telegrapher" goes on to describe the telegrapher's irritations by her customers, which are clearly exacerbated by conflicting gender expectations: "she must therefore, send messages even when they are accompanied by familiar attentions often forced on her by the so-called 'sporting element' found at these public places" and is not allowed to "rightfully resent such acts" because "her wage is dependent on the toleration she shows."

operator, especially nerve-strain—namely, the behaviour of the public." The potential for such problems impelled the Post Office, along with American switchboards, to examine its prospective telephonists for good "nervous equilibrium," besides testing applicants' voices for courtesy and patience.[67] Thus one narrative of nervous delicacy revised another. Women maintained a gentle demeanor more easily than their male counterparts, but the intrinsic sensitivity enabling this feat could itself become a debility on the job. Or as the women themselves might declare, the imperative of continued tact with their customers was enough to drive them as much as it would men to distraction.

However diverse their opinions, fiction writers, business managers, doctors, and operators themselves agreed that communication mediation tasked the sensitive female body; spiritualists maintained the same about their mediums. At first glance, it is easy to read this period's intertwined discussions of technological and occult communications as deriving merely from a similar appearance of eerily immaterial contacts. But beneath the wonder at apparent immateriality were in fact parallel, gendered discussions of materiality—of the receptive feminine bodies in charge of making those contacts possible. Or more precisely, the presumption of women's nervous susceptibility helped to create a middle ground between the immaterial and the material—one conceptually crucial to movements like spiritualism—to the extent that the Victorians commonly used a language of the nerves to explain conditions with assumed, but as yet unspecified, somatic origins.[68] Ultimately the medium's sensitivity might prove both an advantage and a disadvantage, even a disease, with the latter possibility becoming crucial to accounts of the hypnotic mediation of alternate personalities. As I detail in chapter 4, doctors diagnosed the somnambulist's (hypnotic subject's) hysteria much as they had the telephonist's craziness, as a product of a fragile nervous system.

Within this wide-ranging conversation about female sensitivity, the idea of sympathy was liable to grim questions, like the kind implicitly posed in James, about the circumstances this fabled aptitude was obscuring. At the extreme, instead of taking In the Cage as a distortion of the premise of the medium's sympathy, we can see it as a fulfillment of that premise—as a logical exploration of the seeds sown in simultaneously positing women as acutely

---

67. "The Health of Telephone Operators," Lancet 2, no. 2 (1911): 1716; on American switchboard examinations, see Norwood, Labor's Flaming Youth, 41.

68. Oppenheim states that the term functional nervous disorder was used by doctors to describe diseases of the nervous system whose exact physiological foundations were unclear but were expected to be determined eventually ("Shattered Nerves," 8).

receptive and then locating them at the nexus of communication pathways. With her reliance on others' telegrams to bolster her hopes of social relevance, James's heroine offers up the unsettling prospect that there is a fine line between sympathy and eavesdropping. In fact, the medium's sympathy becomes recognizable as simply eavesdropping with its sting quixotically removed: it refashions intellectual interest in vaguer, less threatening terms of emotional affinity. *In the Cage* restores the sense of threat in depicting the medium as someone with a will to interpret the messages that come her way because doing so enables her (for a time at least) to reinterpret herself socially. If one criterion of the ideal medium was that it (she) fade from consideration as a communicative channel, James's telegrapher is, by contrast, exceedingly conspicuous to Everard as well as Lady Bradeen, who is at one point shocked and frightened by her apparent memorization of details of the couple's messages. The sensitivity of the female body was not by itself sufficient to sustain the prospect of intimate human-mediated dialogue. What was also necessary was the disappearance, in effect, of the medium's inquisitive self.

✍ CHAPTER 2

Securing the Line

*Automatism and Cross-Cultural Encounters
in Late Victorian Gothic Fiction*

Equipped with harassing villains, eerie dwellings, sublimely transporting landscapes, and bodily and mental possessions, Gothic literature has always revolved around heightened emotions. Inspiring fear is its stock in trade, and its original eighteenth-century protagonists are creatures of keen sensibilities. A century later, when a spate of new Gothics was published, the mediating woman, with her powers of sympathy, slipped easily into the genre. What she represents in these fictions is a matter of feeling but also, as this chapter explores, a matter of psychical instrumentality. Much of the unease of late Victorian Gothic stems from the specter of the exotic; this character threatens dreadful acts of communication or self-transfer, which the channel often abets, not through her own volition or designs but precisely because these have been overcome. Her function is an unconscious one, and as with her sympathies, it is her femininity that especially enables it. The female channel's recurrence in fin de siècle horror fiction and significance to its affective as well as racial dynamics has yet to be fully acknowledged. Analyzing her role brings out, moreover, how much the menace of another culture is interwoven with issues of communication, the conveyance of knowledge and feeling.

The Gothic trope of the medium rests on Victorian accounts of women's unconscious, or automatic, behavior, accounts with particular purchase around scenes of mediated writing and speaking. As I begin by discussing, the concept of feminine automatism helped to shape optimistic views of how information relays could work, in settings both mystical and mundane. The gendered element in networks framed communications in specific terms—protectedness, insularity, integrity—that would prove, if in unexpected ways, relevant to the plotting of perilous inter-ethnic contacts in the Gothic.

Defending Communications:
The Automatic Woman

When we think of spirit channeling, one of the images that probably comes to mind is of a medium's frantically scribbling hand, as the spirit uses it to write a message. This is a scene of automatic writing, and it reminds us that séance mediumship involved not just a sensitive body but also, in many cases, compromised psychological control of that body. Mediumship helped to inform the Victorians' understanding of automatic behaviors, which entailed a reversion to a partly or wholly unconscious state. Spiritualists held that women, because long on piety and short on will, were particularly well disposed to automatism, offering little impediment to the spirits channeled through them.[1]

But the notion of women's unwitting mediation was not confined to spiritualism. Besides mediumship and related activities rationalized by psychologists, typing and other late Victorian scribal functions often assigned to women became associated with an automatic state of reduced or fragmented attention.[2] Behind this phenomenon in technological mediation, too, seems to have been a gendered logic about personal capacities: the female sex naturally brought to the task a minimum of applied intelligence. About telegraphy, economist Charles H. Garland had this to say in 1901: "Steady work at low pressure, and more or less mechanical in character, necessitating little or no judgment, seems to be admirably performed by women, but where these

---

1. On feminine qualifications for spirit mediumship, see Alex Owen, *The Darkened Room: Women, Power, and Spiritualism in Late Victorian England* (Philadelphia: University of Pennsylvania Press, 1990); Ann Braude, *Radical Spirits: Spiritualism and Women's Rights in Nineteenth-Century America* (Boston: Beacon, 1989); R. Laurence Moore, *In Search of White Crows: Spiritualism, Parapsychology, and American Culture* (New York: Oxford University Press, 1977), chap. 4; Vieda Skultans, "Mediums, Controls, and Eminent Men," in *Women's Religious Experience: Cross-Cultural Perspectives,* ed. Pat Holden, 15–26 (London: Croom Helm, 1983); and Jennifer A. Yeager, "Opportunities and Limitations: Female Spiritual Practice in Nineteenth-Century America," *ATQ* 7 (1993): 217–28.

2. Lisa Gitelman, *Scripts, Grooves, and Writing Machines: Representing Technology in the Edison Era* (Stanford: Stanford University Press, 1999), chap. 5.

conditions are lacking they are generally found inferior to men."[3] Analogous assumptions about typing prevailed enough for a group of women in the profession to protest in 1902 against the "alleged mechanical nature of the work we perform" and to affirm that "every person of experience ... knows that the efficient performance of the duty is impossible without the possession of average judgment and skill."[4]

The insidious consistency of the trope of automatism/mechanism in discussions of women media suggests its centrality to perceptions of mediation itself and provides a key for understanding in more detail the reasons behind women's prevalence in the vocation. Women seemed naturally made for typing, Christopher Keep argues, because the popular blind or touch method, which regulated bodily impulses and prevented the typist's self-centered gaze at her own work product, jibed with the view that women lacked subjective agency, a view evident around automatic writing as well.[5] That raises the question: how did this lack of agency matter—what were its perceived benefits—in terms of the most obvious goal of the process in which the typing or ghost medium figured, communication? Late Victorians like Henry James were uncomfortably aware that technologically mediated dialogues, like spiritualistic ones, were really no longer dialogues strictly speaking because they always entailed a third party potentially knowledgeable about the information transmitted. In the context of this drawback of emerging communication methods, I propose, people in all manner of scenarios wanted to insist on go-betweens as bodies without consciousness. Common beliefs about women's limited capacity for "brain-work" made it simple to think of them as just such weak-minded conduits, and those representations assuaged

3. Charles H. Garland, "Women as Telegraphists," *Economic Journal: The Journal of the British Economic Association* 11 (1901): 258.

4. Quoted in Meta Zimmeck, "Jobs for the Girls: The Expansion of Clerical Work for Women, 1850–1914," in *Unequal Opportunities: Women's Employment in England 1800–1918,* ed. Angela V. John (Oxford: Blackwell, 1986), 168. In another essay, Zimmeck notes that in pre–World War I jobs at the Civil Service, both male hand-copying and female typing were designated as "mechanical" (as opposed to other departmental jobs dubbed "intellectual"); "'The Mysteries of the Typewriter': Technology and Gender in the British Civil Service, 1870–1914," in *Women Workers and Technological Change in Europe in the Nineteenth and Twentieth Centuries,* ed. Gertjan de Groot and Marlou Schrover (London: Taylor and Francis, 1995), 88. But this does not diminish that women were welcomed into many of the posts thus designated; in other words, we should still ask what made women seem right for the growing field of "mechanical" word processing.

5. Christopher Keep, "Blinded by the Type: Gender and Information Technology at the Turn of the Century," *Nineteenth-Century Contexts* 23 (2001): 149–73. Others have queried the textual medium's real level of self-subtraction, noting that dictated writing complicates the locus of authorship: see Gitelman, chap. 6, and Bette London, *Writing Double: Women's Literary Partnerships* (Ithaca: Cornell University Press, 1999), esp. chaps. 5 and 6.

fears about eavesdropping, among other distortions of knowledge transmission, by emptying out the medium's interposed presence.[6]

Michèle Martin shows that the concern for privacy was one factor underlying the preference for women at the telephone switchboard. To some extent, the operator had to listen in on callers' conversations in order to keep track of charges and disconnections; and yet her employers, trusting in female submissiveness and vigilant about their subscribers' right to privacy, demanded that the operator be largely mentally absent, engaging only in "*civil* listening, which implied that she ought to attempt to hear only the 'sounds' of the callers' voices and to ignore the meaning.... This is a typical case in which the operator was not expected to act as a human being, but to identify herself with an object, a machine."[7] That the telephone industry would have this expectation indicates not only its apprehensions about having a live person at the switchboard but also how the fiction of women's easy reversion to a mechanical state could solve the problem.[8] But it is important to reiterate that such fictions, as a foundation for ideas of women's usefulness, were sweeping ones: fictions, in other words, not necessarily contingent upon the internal workings of any single industry or business, such as, in Martin's analysis, Bell Telephone. What an occupation-specific theory of feminization cannot finally account for is the sheer diversity of fields of female mediation, nor can it account for the reputation for effortless mechanism or automatism that at one time or another hung about every one of them.

During the Victorian era, there rose to prominence, due to its existence in many different forms and many different settings, a certain kind of communication—the human-mediated communication—and that communication was itself prone to being feminized. In its ideal form, the medium was an inherently feminine thing—unobtrusive, non-interfering, effectively

---

6. On Victorian truisms about women's deficiencies as "brain-workers," see, e.g., Elizabeth K. Helsinger et al., *The Woman Question: Society and Literature in Britain and America, 1837–1889*, vol. 2, *Social Issues* (Chicago: University of Chicago Press, 1983), 75–89.

7. Michèle Martin, *"Hello, Central?": Gender, Technology, and Culture in the Formation of Telephone Systems* (Montreal: McGill-Queen's University Press, 1991), 69. Martin's study deals with the Canadian telephone industry, but because Bell Telephone was Canada's reigning service provider, her observations about management can be applied to many American switchboards, if not also to other locales emulating Canadian and American policies.

8. At times the assumption of women's self-effacing behavior at the switchboard can seem willfully naive. Operators did eavesdrop, as Martin records (*"Hello, Central?"* 107–8). In fact, women could seem to present a special problem when it came to eavesdropping: originally some telephone managers balked at hiring them, presuming they would gossip; Stephen H. Norwood, *Labor's Flaming Youth: Telephone Operators and Worker Militancy, 1878–1923* (Urbana: University of Illinois Press, 1990), 280. Yet in one of the many convolutions within this era's discussions of mediating women, one gendered cliché may well have extenuated the other—the notion of women's better ability for practices like "civil listening" abating concerns about their chattiness.

nonexistent for the purposes of intellectual exchange. The label of "automatic" is a first clue to this perception even while, as a description of the female employee, it was not exclusive to the arena of office or telecommunications work. That is, the premise of the unthinking woman worker took on an added utility within the latter domain: it was not merely the case, as might be true in the turn-of-the-century factory, that the rhetoric of automatism worked to assure people of the efficiency of processes of production.[9] It also assured them of the integrity or value of the product itself; it indicated that dialogue or information would not be *de*valued in being either exposed to or corrupted by the intervening medium.[10]

The connection between mechanical women, private communication exchange, and the value of the communication is also evident in settings involving the paranormal. Séances might offer sitters anodyne descriptions of the hereafter, but these alone did nothing to prove the legitimacy of the messages or of the medium purporting to channel them. The private or personal nature of the information channeled was a gold standard. When in the late 1880s the medium Mrs. Leonora Piper convinced Richard Hodgson and other psychical researchers of her talents, she did so by delivering up personal data about her séance attendees; this seemed to prove real occult power, telepathic or perhaps even spiritualistic.[11] But there was an important concomitant measure of veracity at work for the spirit medium: not only must she convey information so private only a deceased intimate of the sitter could know it; she must at the same time convey it by some other source than her own intelligence. It was this second criterion Hodgson was thinking of when he had Mrs. Piper followed by detectives to ensure she had no secret access to her sitters' lives.[12]

---

9. Contrast Keep, who reads the fantastical picture of the woman typist as an "enabling absence" in terms of a desire for speedy output ("Blinded by the Type," 159). On feminine automata versus masculine productive power in the factory, see M. Norton Wise, "The Gender of Automata in Victorian Britain," in *Genesis Redux: Essays in the History and Philosophy of Artificial Life,* ed. Jessica Riskin, 163–95 (Chicago: University of Chicago Press, 2007).

10. My thanks to Dan Seward for conversations that helped me to formulate the latter point.

11. Janet Oppenheim, *The Other World: Spiritualism and Psychical Research in England, 1850–1914* (Cambridge: Cambridge University Press, 1985), 373. John Durham Peters notes, too, the importance of the "private, contingent fact" in validations of Mrs. Piper's mediumship, *Speaking into the Air: A History of the Idea of Communication* (Chicago: University of Chicago Press, 1999), 190. Also on Mrs. Piper's dealings with psychical researchers, see Ruth Brandon, *The Spiritualists: The Passion for the Occult in the Nineteenth and Twentieth Centuries* (New York: Knopf, 1983), 232–35; and Alan Gauld, *The Founders of Psychical Research* (New York: Schocken, 1968), 251–74. Predictably, her eventual confession of fraud, published under a blaring headline in the *New York Herald,* articulated her mediumship in technological terms: "I AM NO TELEPHONE TO THE SPIRIT WORLD" (quoted in Brandon, *Spiritualists,* 233).

12. Oppenheim, *Other World,* 374.

Florence Marryat's *The Dead Man's Message* (1894) implies similar tests of veracity for the spirit medium in the novel, Mrs. Blewitt. A young woman named Maddy seeking communications from the dead visits Mrs. Blewitt, who channels Maddy's mother; but Maddy remains unsure that it is really her mother doing the communicating: "O, I remember you, and I love you...; at least, I love my mother's memory; but how am I to know that you are she? Give me some proof. Tell me something that no one can know but our two selves—that Mrs. Blewitt is not aware of."[13] Maddy's last statement links the two necessities for validating the spirit message: it must touch upon personal matters of which the medium herself, however, has no ordinary knowledge. Even in those cases when the medium was not entranced but instead remained cognizant of the scene around her, she was still presumed to be writing "automatically" or ignorantly, in some basic sense unwillfully, of the intimacies she transmitted. In spiritualism as in human-mediated communications generally, salvaging the illusion of private dialogue—and by extension, in spiritualism's case, of genuineness—required asserting the medium's automatism.

The automatized woman had no *ordinary* knowledge of what she transmitted: the distinction was crucial in Victorian and early twentieth-century understandings of mental processes. Automatism did not altogether deny knowledge; it simply split it, conceptually speaking, into conscious and unconscious forms. To perform automatically was to perform unconsciously—that is, to have one's body execute actions of which the conscious mind was only incompletely if at all aware or in control. This became coded as a feminine capability, as did sympathy or sensitivity, with which automatism was supposed to cooperate in the person of the medium. Further, in medical and other discourse, unconscious behaviors and sensitivity were both often associated with nervous delicacy.[14] Hence accentuating woman's nervousness could serve all in one stroke as a means of reducing her to a feeling body more or less evacuated of consciousness. To be intellectually withdrawn but sensitively present: this was what women seemed capable of achieving as communication media. And in both respects—this particular kind of withdrawal and this particular kind of presence—women's participation encouraged the fantasy that the human beings within mediated communications could recreate the seamless, in some cases even intimate interconnections of non-mediated dialogue. To be sure, that last statement suggests a perfect synchrony of qualities, automatism and sensitivity, that in fact hardly

---

13. Florence Marryat, *The Dead Man's Message* (1894; New York: Arno, 1976), 125.
14. My fourth chapter elaborates on this nexus in looking at a case study by Dr. Morton Prince.

characterizes every representation of mediating women, in fiction or else-where.[15] I nonetheless emphasize these qualities together because both so often circulate around such discussions and because, as I have indicated, there is an inherent complementarity between the two, a complementarity that becomes most noticeable when we look at individual figurations in light of one another instead of in isolation.

Looking at a variety of representations, we also see how much Victorian and later writers tended to conceptually blur diverse sorts of mediated communication—hypnosis with typing, spiritualism with telegraphy, and so on. The tendency is evident even in writings that do not concern communication as such. Take for example theories of the crowd, which held that the demagogue's captivating rhetoric overwhelms the personal will of individual crowd members and produces an unconscious mind-meld among them instead. These ideas obviously took their hue from occultist renderings of psychical processes. The crowd's collective mental state suggests telepathy; and a work like Gustave Le Bon's *The Crowd* (translated into English in 1896) draws expressly on ideas of mesmeric suggestion, stressing the persuasion of the demagogue who leaves the crowd "hypnotised" by his "magical phrases" and "words and formulas," the "very vagueness" of which "wraps them in obscurity" and "augments their mysterious power."[16] Then again, given the language of electromagnetism common to occult and technological discourse, one cannot always be sure whether *The Crowd* is invoking a mass mesmerism or a telegraphic relay. Which does Le Bon mean to signify when he declares of the demagogue's speech that certain "transitory images are attached to certain words: the word is merely as it were the button of an electric bell that calls them up"?[17] The crowd resembles both at once in being a smooth medium of ideas between its members and its leader, and between members themselves. Significantly, Le Bon interprets this medium

15. Nor is it the case that the stereotype of mindlessness plagued all women in mediating positions all the time. Venus Green notes, for instance, that in the earliest years of the Bell System, operators had a relatively familiar relationship with subscribers and were expected to please them by performing socially and emotionally through a combination of "[p]ersonality and intelligence." *Race on the Line: Gender, Labor, and Technology in the Bell System, 1880–1980* (Durham: Duke University Press, 2001), 19. In any given vocation, clearly, various goals of communication surrounded mediation and might be differently prioritized at different times. I am arguing that the notion of inferior female mental power was available when particularly needed to address concerns about the transmission of knowledge.

16. Gustave Le Bon, *The Crowd: A Study of the Popular Mind* (1896; Atlanta: Cherokee, 1982), 26, 182–83, 96–97. On the paranormal and especially telepathic premises of crowd theory, see also Pamela Thurschwell, *Literature, Technology and Magical Thinking, 1880–1920* (Cambridge: Cambridge University Press, 2001), 27–28.

17. Le Bon, *Crowd*, 97.

as intrinsically feminine: "Crowds are everywhere distinguished by feminine characteristics," which translate as a tendency to "simplicity and exaggeration of the sentiments."[18] With its simple-mindedness and emotional impressionability, the crowd, it would seem, is simply the female medium writ large and mapped onto the structures of modern society.

Instant Oriental Messaging

To fully register how the medium's exaggerated sentiments could signify within the Gothic, we need first to appreciate that this was a time when Victorians were exploring the affective and informational uses of communication methods in forging cultural identity and in intercultural conflicts. In an 1893 article for the *Contemporary Review*, MP J. Henniker Heaton reasoned that making international telegraphy affordable was crucial to maintaining imperial integrity because it would allow people in England to converse with colonial settlers, preventing the "strain of divergent sympathies and interests." Though Britain's domestic telegraph had been nationalized, international submarine cables were in the hands of private companies; Heaton lamented that because of these companies' overcharging,

> not one of the many millions of humble and honest toilers at the Antipodes has been able to cable the "old folks" in the Mother Country one word of intelligence or sympathy... A beneficent invention, the common heritage of our race—one that might enable all Christendom to assemble as it were under one roof, there to talk, laugh, and weep together—has been selfishly appropriated by a few speculators.[19]

The telegraph reminded dispersed citizens of their shared sentimental investment in Great Britain within a homey, feminized emotional space: allied with the "Mother Country" "under one roof."

The telegraph was, moreover, essential to exercising military authority within the colonies. Where some of Heaton's contemporaries might see the

---

18. Le Bon, *Crowd*, 20, 33.

19. J. Henniker Heaton, "Imperial Telegraph System: Cabling to India and Australia," *Contemporary Review* 63 (1893): 537, 543. Heaton mimics a logic expressed by Victorians like journalist W. T. Stead, who Roger Luckhurst argues had manifold hopes of uniting people in mental harmony and who imagined communication technologies as instruments for creating emotional connections across the Empire. *The Invention of Telepathy: 1870–1901* (Oxford: Oxford University Press, 2002), chap. 3.

so-called nerves of the telegraph network as conduits for the kind of affection he theorizes will reunite "our race," he imagines them as agents of precise sensorimotor power:

> The analogy between the imperial cable system and the human nervous system must have struck everybody. By means of electricity, it is as easy to command and concentrate on a given point the diffused strength of a dozen great nationalities as it is for a man to employ eye, foot, and hand together in delivering a crushing blow. Every moment orders, inquiries, reports, and advice are being flashed between Whitehall and British officials in all parts of the globe. The cable multiples the strength of our fleets and armies, and is an essential part of our governmental machinery.[20]

Besides bolstering national strength by linking those who defined its ethnic identity, the telegraph could help to control and even, Heaton's analogy menacingly suggests, deliver a "crushing blow" when needed to those who did not. Both of these uses of the telegraph were made clear during the 1857 Indian Mutiny. On the one hand, preachers in England declared it a conduit for rousing the hearts of the nation in sorrow over reported brutalities against Anglo-Indians.[21] On the other hand, on the ground in India, wired communications helped the British to track their opponents' movements and arrange troops rapidly, making telegraph lines a repeated strategic target for destruction by the insurrectionists.[22]

By some accounts, though, the Mutiny set the stage for more striking feats of communication than telegraphy. In the preface to his primer *Wireless Telegraphy, Popularly Explained* (1903), Richard Kerr asserts that the Indians countered British telegrams with their own stealthy conveyances of information. In addition to transporting bread stamped with secret messages, the rebels communicated with each other "in more occult ways that were harassing to the military authorities, but the secret was not discovered." Kerr describes similar troubles in the war in Afghanistan, where a British mission fell through because the locals got wind through undetectable means of planned military movements. He continues with some awestruck comments

---

20. Heaton, "Imperial Telegraph," 538.

21. Don Randall, "Autumn 1857: The Making of the Indian 'Mutiny,'" *Victorian Literature and Culture* 31 (2003): 13.

22. Saroj Ghose, "Commercial Needs and Military Necessities: The Telegraph in India," in *Technology and the Raj: Western Technology and Technical Transfers to India, 1700–1947*, ed. Roy MacLeod and Deepak Kamur, 153–76 (New Delhi: Sage, 1995).

about clairvoyance and telepathy, deducing, "It would seem that the Oriental methods of signalling without wires must rest entirely on a highly trained mental effort. . . . It is quite possible that the Orientals give more attention to this kind of mental training, and are therefore able to accomplish the wonderful feats of signalling attributed to them."[23] Kerr may sound like a crackpot, but his ideas about the "Oriental" talent for telepathy, as well as its application during military conflicts, appear in other sources as well. For example, in an 1894 essay in *Arena,* "The Wonders of Hindoo Magic," Heinrich Hensoldt writes of telepathy as an ancient and quotidian practice in India. In another article on "Occult Science in Thibet" that same year, Hensoldt roughly anticipates Kerr in stating that "it invariably happened that the news of any success or disaster to the British [during the Afghan war] was known all over India long before the authorities at Calcutta were officially informed," notwithstanding the government's benefit of telegraphic couriers.[24]

The notion of the Oriental's occult prowess repeatedly shapes the intercultural dramas of fin de siècle horror fictions; my aim is to highlight the importance of the mediating woman and her magically manipulable automatism in these plots. For Daniel Pick, George Du Maurier's *Trilby* (1894) is symptomatic both of a late Victorian association of mesmerism and occultism with Jewishness, Asia, and the Middle East, and of fictional monsters who "embody a kind of danger which ultimately colonises their prey from *within*."[25] What needs now to be recognized is that women provide some of the most spectacular of these prey, in part because they possess a reservoir of feeling and morality that adds to the gravity of the cultural contest by accentuating its potentially ghastly consequences: polluted personal, ethnic, and bodily sympathies. As also for Heaton, the strength of British culture relies on the strength of emotional tendencies, figured in terms of the feminine—and, in these novels, as capable of being perverted or blunted. Women's susceptible

23. Richard Kerr, *Wireless Telegraphy, Popularly Explained* (London: Seeley, 1903), 2, 6–7.

24. Heinrich Hensoldt, "The Wonders of Hindoo Magic," *Arena* 10 (1894): 46–60; Hensoldt, "Occult Science in Thibet," *Arena* 10 (1894): 370.

25. Daniel Pick, *Svengali's Web: The Alien Enchanter in Modern Culture* (New Haven: Yale University Press, 2000), 222; see also 132. Luckhurst claims that the late Victorian Gothic is significantly shaped by conceptions of trance and remote influence (*Invention of Telepathy,* chap. 5, esp. 204–13); my project here is to focus more extensively on the function of femininity within the kinds of texts he discusses. For a different angle on trance and gender, see Sarah Willburn's argument that authors ranging from Charlotte Brontë to Marie Corelli use entranced heroines to envision new possibilities for civic community, diverging from typical Victorian social philosophies of possessive individualism and gendered separate spheres; *Possessed Victorians: Extra Spheres in Nineteenth-Century Mystical Writings* (Aldershot, UK: Ashgate, 2006), chap. 4.

impressionability according to the discourse around trance operates as a generic crux, making the exotically controlled female a main locus of fear and suspense.

The role of the mediating woman within the late Victorian Gothic turns not just on her sympathy but also on her automatism. A master of occult communications, the Oriental can initiate feminine trance and encroach on Occidental culture through it, transforming the heroine into figuratively or literally disputed ethnic terrain. Positing the automatic woman as an occult portal for Oriental predators, these fictions thus take advantage of an irony implicit in automatism on the turn-of-the-century communications scene generally, one rooted in its dual nature. On the one hand, automatism protects the privacy or security of others' talk from the medium's own consciousness, from her claims to knowledge, restoring the insularity of dialogue. But on the other hand, in novelistic landscapes, it compromises security and insularity by converting her into an entryway for sometimes menacing forces and entities who operate *un*consciously. By splitting the medium's mind into conscious and unconscious forms, then, the trance state patches potential leaks in the socio-communicative network and creates them simultaneously.

All this is not to say, however, that the concept of automatism is monolithic as an indication within narrative of female capacities. A text like *Trilby* does not much question the verisimilitude of a woman serving as facile machine of another. By contrast, Bram Stoker's *Dracula* (1897), like the writings of bestseller Marie Corelli, depicts the automatic woman's communicative role to reimagine the reputed relationship between gender and brain power.

Warring Communications, Contested Sympathies

The bite of Stoker's Count—fundamental to his project of vampiric self-diffusion—establishes blood ties that are also communicative ties. In preying on Mina Harker, Dracula generates a telepathic connection that can unconsciously apprise him of all she knows about his pursuers' movements, a possibility Van Helsing understands well enough to guard against: "we must keep her ignorant of our intent, and so she cannot tell what she know not."[26] Vampirism is an Eastern talent. Jonathan's journal entries conflate Dracula's

---

26. Bram Stoker, *Dracula,* ed. Maud Ellmann (1897; Oxford: Oxford University Press, 1996), 323. All further page references to *Dracula* are parenthetical within the text.

homeland with the Orient.[27] And when Van Helsing, trying to convince Seward of wonders unexplained by science, expounds on the Indian fakir, who at his death is buried in a sealed grave but, when "men come and take away the unbroken seal," lies "not dead...but...rise up and walk amongst them as before," he could equally be describing the un-death of Dracula and his follower Lucy, who likewise escapes from a locked grave (192). Dracula's telepathic proficiency is a product of his Oriental origin and an ethnically specific avenue of exchanging knowledge and exerting control. If Dracula's foray into England constitutes a colonial project that mirrors Victorian colonialist efforts,[28] then he is using telepathy in the same way the British have learned to use telegraphy: as a weapon to institute lines of communication that advance or maintain a colonialist mission.

Mina's unconscious channeling is a treacherous doubling of her work for Van Helsing and his small army. The group depends on progressive technologies—shorthand, the phonograph, the telegram—to track the Count's activities, and as a typist Mina transcribes their records, making them available for collaborative analysis. The vampire and his opponents are thus engaged in a war of culturally disparate means of communication, and as both dual medium and potential prey, Mina is the focus of this war.[29] Certainly Dracula is not restricted to occult transmissions; he is perfectly capable of using English instruments and bureaucracies for his own ends. Nonetheless, his unconscious link to Mina is surely the most striking of his methods and principally defines his strategy as the narrative progresses. In the end Dracula is defeated—and yet his communications show up his opponents' methods in a manner that ultimately casts doubt on their ways of understanding and operating within the world. What is more, this is a lesson about the limits of knowledge that encompasses more than just an awareness of vampirism. Concentrating on the communications war and Mina's implication in it allows us to grasp her

---

27. Jonathan reflects, "It seems to me that the further East you go the more unpunctual are the trains. What ought they to be in China?" (2); later he observes that the locals resemble an "old Oriental band of brigands" (3).

28. See Stephen Arata, *Fictions of Loss in the Victorian Fin de Siècle* (Cambridge: Cambridge University Press, 1996), chap. 5.

29. Jennifer Wicke likens *Dracula's* vampires to its technologies, relating vampiric to mass cultural consumption; "Vampiric Typewriting: *Dracula* and Its Media," *ELH* 59 (1992): 467–93. Laura Otis explores the novel's abundant, sometimes overlapping technological, occult, and organic networks of communication; *Networking: Communicating with Bodies and Machines in the Nineteenth Century* (Ann Arbor: University of Michigan Press, 2001), chap. 6. My own emphasis is less on the similarities than on the differences between Dracula's and his foes' methods for transferring knowledge—particularly on the important Orientalization of telepathy—and in turn on the communications conflict the novel stages.

group's reeducation about the conditions of belief, and to see the novel as doubly a tale about their path to a belief in the unbelievable: in the existence of the vampire, of course, but also in the strong, as opposed to facilely automatic, female mind.

Initially, though, we should note that Dracula and his foes struggle over not only Mina's role as information nexus, but also the orientation of her sympathies. By nature Mina is the soul of feminine compassion, as we see in her response to dead Lucy's suitors. Soon after meeting Seward and listening to his phonographic diary, Mina enters his study with eyes "flushed with crying" to learn of his pain over her friend (222). This is the same "sympathy" she later shows to the grieving Arthur (230) and to Quincey. Mina's quick warmhearted impulses toward these men, whom she has only just met, make her, even more than Lucy, the emotional pivot of the group, grounding the allied defense of England in affective bonds. But that only underlines the tragedy as well as triumph of Dracula's rapport, as it reforms her sentiments in monstrous ways that favor her friends' enemy. The reformation is most vividly demonstrated during the final days of the pursuit. As she and Van Helsing near the Count's castle, the professor records her eerie mirth: "all at once I find her and myself in dark; so I look round, and find that the sun have gone down. Madam Mina laugh, and I turn and look at her. . . . I am amaze, and not at ease then" (364). When the trio of she-vampires approaches, Van Helsing realizes Mina shares none of his anxiety for her well-being: "I looked at her, but she sat calm, and smiled at me . . . I turned to her, and looking in her eyes, said:—. . . 'It is for you that I fear!' whereat she laughed—a laugh low and unreal, and said:—'Fear for *me*! Why fear for me? None safer in all the world from them than I am'" (367).

Mina's laughter, like her lack of fear, is a principal signifier of the redirection of her affective loyalties toward the Count's minions. It is also a sinister echo of a moment in Grant Allen and May Cotes's *Kalee's Shrine* (1886), another imagining of Oriental insinuation through the unconscious female body. In the prologue, set in India, the English two-year-old Olga Trevelyan is brought by her ayah to a hidden shrine for a secret consecration to the "Black One," Kali or "Kalee," a destructive Hindu goddess worshiped by murderous Indians. The priest makes a ritual incision with a flint at each of Olga's temples that he says will keep her "spellbound" by night to Kalee, even while she remains a Christian by day.[30] The rest of the text takes place at a house party in an English coastal town—where the hostess, lamenting

30. Grant Allen [and May Cotes], *Kalee's Shrine* (1886; New York: New Amsterdam Book Company, 1897), 11. All further page references to *Kalee's Shrine* are parenthetical within the text.

the redundancy of single English women due to the exportation of men to the colonies, advocates a technological solution for generating affections: a "Universal British Empire Telephonic Matrimonial Agency, to bring the young people everywhere together" (85). Meanwhile, Olga, now eighteen and a guest at the party, begins to exhibit signs that the priest's ritualistic operation has succeeded. In a chapter whose title, "Kalee in Suffolk," forebodes the infiltration of the exotic, Alan Tennant, who is just beginning to woo Olga, is shocked to spy her gazing from her bedroom window at the wreck of an incoming ship with "no tinge of sympathy or suspense or terror. . . . She laughed like a maniac at the horrible catastrophe; laughed, and laughed again, with inextinguishable merriment, as though the sight . . . were to her unnatural soul the most amusing and delightful episode in all creation" (44–45).

The conflict between India and England plays itself out on Olga's body, specifically at her eyes. The emphasis on the devotional significance of the incisions around her eyes reinforces the double entendre of *temple* and the service her body may provide for Kalee. Indeed, Olga herself may be the shrine of the book's title. But as a scientist, an oculist in particular, Alan refuses to grant the premise of Kalee's furtive presence through Olga. He interprets her inability to close her eyes during sleep, an apparent sign of her devotion, as merely the result of the Indian priest's deliberate severing of the nerves at her temples. The severed nerves give, symbolically, a physical cause for the insensitivity Olga exhibits for those on the sinking ship. Moreover, they are a site for refining the book's conflict around the theme of vision as one between Indian superstition and English science. When one evening Olga begins speaking unconsciously in Hindustani and later in her sleep tries to murder a friend, Sir Donald Mackinnon, a Scotchman himself prone to "pure Highland superstitiousness and nonsense" (234) and a former Indian civil servant, is certain that Olga has acted at the behest of Kalee. But Alan, who has already rationalized Olga's hardhearted behavior at the shipwreck as a case of somnambulism, offers another explanation: it is Sir Donald's own incessant talk of Kalee, coupled with Olga's entrancement by a mesmerist who provided the evening's entertainment, that has worked on her "delicate nervous organization" and unconsciously prompted her violent action (208). Alan's final maneuver to rid Olga of the fiction of Kalee is to undo the mischief of the priest who prevented her from closing her eyes by making his own cuts with "dexterous gentleness" at her temples (229).

In the most literal manner, then, Olga's nervous body is wrested away from exotic myth and placed under the reparative authority of English medicine. At the end of the novel, nonetheless, the reader is left with an uncertain sense of just where myth has ended and science has begun. In declaring

Olga's unconscious response to Sir Donald's indirect suggestions as a case of "hysterical somnambulism" (208), Alan gives his scientific imprimatur to a phenomenon that looks much like the psychical action at a distance he has deemed fantastical. The reality of the latter is dismissed only up to the point where Alan begins to conceive of it not as through a window onto Indian culture, but as in a mirror that reflects what his own culture says is true and possible, namely the mesmeric manipulation of another.

Alan's recasting of Olga's trance indicates the shifty status of mesmerism for the Victorians. Despite our likely assumptions that it stood categorically outside the realm of valid science, matters were actually far from being this clear. In the high Victorian period, mesmerism did not so much confront as help to define scientific authority in providing scenarios for playing out power relationships and debating the contours and conditions of acceptable knowledge.[31] During the late nineteenth century, it occupied an especially ambiguous position between legitimate knowledge and its impostor thanks to psychical research. Next to, or rather together with, theories of the unconscious/subliminal mind posited by prominent figures like Frederic Myers, hypnotism—which for many writers, like Corelli, was simply mesmerism or "animal magnetism called by a new name"[32]—was arguably the most crucial aspect of psychical research to confuse the boundaries of science, linking as it did the interests of occultists with those of reputable psychologists. *Kalee's Shrine* demonstrates a second, related shiftiness, in that the phenomenon of psychical influence has a place in the cultures of both colonizer and colonized. It is only Occidental and scientific bias that prevents Alan from recognizing the proximity of his practice to Oriental enchantment and hence the possible credibility of both.[33]

Late Victorian Gothic fiction exploits this dually fraught categorization of trance influence. By presenting psychical operations on the automatic woman as available for ostensibly opposing cultural spheres of belief, a work like

---

31. See Alison Winter, *Mesmerized: Powers of Mind in Victorian Britain* (Chicago: University of Chicago Press, 1998).

32. Marie Corelli, *A Romance of Two Worlds* (1886; Alhambra, CA: Borden, 1986), xx. All further page references to *A Romance of Two Worlds* are parenthetical within the text.

33. Alan's blind spot corresponds to the views of actual authorities in the Victorian colonies. Winter describes British surgeon James Esdaile, who used mesmerism as an anesthetic in colonial India and in so doing mimicked the practices of local healers, while also purportedly offering a needed rational framework for understanding them. Thus, as Winter finely sums up, mesmerism "had to be exported to Britain and brought back to India in order to become knowledge. Yet the process of proving the validity of this knowledge involved a return to the world of magic and wonder from which the excursion through Western science had been intended to distinguish it" (*Mesmerized*, 211).

Kalee's Shrine highlights the practice of hypnotism as a steadfastly uncertain ethnic, epistemic, and moral ground. In *Dracula,* too, hypnotism is a no-man's-land. "I suppose now," Van Helsing taunts the dyed-in-the-wool scientist Seward, "you do not believe in corporeal transference. No? Nor in material-ization. No? Nor in astral bodies. No? Nor in the reading of thought? No? Nor in hypnotism—" But hypnotism is the exception, because, as Seward replies, "Charcot has proved that pretty well" (191). Hypnotism provides the common ground between Western science—Jean-Martin Charcot's research on hysterics at the Salpêtrière Hospital in Paris—and its occult challenges. The hypnotic channel Van Helsing eventually creates in Mina as a way of tracking the vampire is in basic form little distinguishable from Dracula's telepathy. Indeed, the professor talks of the two techniques as one operation simply flipped on its head: "If it be that she can, by our hypnotic trance, tell what the Count see and hear, is it not more true that he who have hypno-tize her first... should, if he will, compel her mind to disclose to him that which she know?" (323). Even so, hypnotism and the Western neurological theories it was often allied with will mark out key distinctions between Van Helsing's designs and Dracula's. Although Van Helsing considers himself to have surpassed Seward in broadmindedness, his perspective does not stray far from Charcot's vision of hysterical femininity, with telling consequences for the communications war.

The Myth of Mina's Nerves

Early in the hunt, it is difficult for Mina's male allies to see her much differ-ently than Alan Tennant sees Olga Trevelyan, as more than an impressionable and nervously delicate creature. Van Helsing decides that in the battle against Dracula there can be "no part for a woman... hereafter she may suffer—both in waking, from her nerves, and in sleep, from her dreams" (235), a sentiment seconded by Seward and later parroted by Jonathan, who fears seeing his wife's "nerve broken" (262). Such statements are premised on the Victorian conception of women's weaker nerve force and hence inability to withstand mental and emotional strain; medicine interpreted women's smaller crania and lower supply of neural energy as restrictions on their cerebral output—hence, supposedly, the dearth of women geniuses.[34] The irony in the case of *Dracula* is that Mina is so resolute and arguably the most

---

34. On these presumptions beneath the evaluations of Mina, see John L. Greenway, "Seward's Folly: *Dracula* as a Critique of 'Normal Science,'" *Stanford Literature Review* 3 (1986): 220–21.

discerning member of the group, originating the crucial strategy of collating the group's various records into a coherent narrative and producing multiple copies for collaborative analysis.[35] Mina's nerves may well be responsible for her receptivity to her allies' grief.[36] Yet they are also strong enough to tinge that feminine sympathy with a powerful inquisitive drive. She may cry with Seward, but she still insists on publicizing his phonographic diary of the events surrounding Lucy's death—"we must have all the knowledge and all the help which we can get" (222)—and her insistence, adorned by bravery as much as charm, is impossible for Seward to refute: "She looked at me so appealingly, and at the same time manifested such courage and resolution in her bearing, that I gave in at once to her wishes" (223).

Jonathan's comments on his wife's weak nerves are especially ironic given that he himself has suffered a "violent brain fever" after his experience at the Count's castle and, moreover, that Mina has sought to protect him from further strain by taking on his traumatic knowledge (99). "I am always anxious about Jonathan, for I fear that some nervous fit may upset him again," she notes in her journal (171–72). Indeed, Jonathan practically invites her protection in giving her his journal: "I have had a great shock, and when I try to think of what it is I feel my head spin round...The secret is here, and I do not want to know it....Take it and keep it, read it if you will, but never let me know" (104). Once Mina reads his record, after Jonathan, still nervously exhausted, passes out when he sees the Count in Piccadilly, she immediately recognizes its potential usefulness as evidence: "I shall get my typewriter this very hour and begin transcribing. Then we shall be ready for other eyes if required. And if it be wanted, then, perhaps, if I am ready, poor Jonathan may not be upset, for I can speak for him and never let him be troubled or worried with it all" (179).

There are two crucial points to be made here. First, Mina's demonstration of mental fortitude again proves inseparable from her secretarial work. Just as when she transcribes Seward's phonographic cylinders and the group's other records, her typing of Jonathan's journal entails a carefully thought out processing of knowledge. Second, this analytical component of Mina's function as information channel eventually becomes impossible for her allies to miss. Criticism on *Dracula* has often focused on the incongruity between

---

35. "Let me write this all out now. We must be ready for Dr Van Helsing when he comes....In this matter dates are everything, and I think if we get all our material ready, and have every item put in chronological order, we shall have done much....Lord Godalming and Mr Morris are coming too. Let us be able to tell them when they come" (224).

36. Otis suggests that the novel implies the scientific judgment of women's nervous sensitivity in the female characters' sympathy and unconscious openness to Dracula (*Networking,* 212–14).

Mina's mental strength and her allies' unwillingness, as in their quick dismissal of her from the hunt, to grant her this trait. One resolution of the incongruity has been to read her intelligence as one of the many ideological threats evoked by the vampire that the novel ultimately quashes: the New Womanish "man's brain," as Van Helsing dubs Mina's, is a distraction from nurturing priorities that are reasserted with the image of her as mother in the story's postscript. But on the contrary, I would propose that the novel progresses as a claim rather than a disclaiming of Mina's intellect. That is, the contradiction between what she is and what her allies perceive her to be is only an initial error from which they are eventually schooled away, largely by experience of her unusually incisive work as a communication medium.

If the men try to keep Mina ignorant not once but twice, we should note that only the first of these attempts is based on the false premise of her nervous delicacy. The second is a defense against Dracula's telepathy and is in fact sound strategy. For the vampire's telepathy is a powerful weapon in the communications war, outstripping his opponents' vengefully up-to-date methods, most notably their own use of trance. Although hypnotism is available in *Dracula* to both Western science and Eastern magic, it proceeds through different techniques, and with very different results, for the two sides, in a way that amplifies the nagging doubt about the superiority of scientific perspectives prompted by a tale like *Kalee's Shrine*. These two aspects of the communications war in *Dracula*—Mina's intelligent channeling for her allies and Dracula's greater facility with automatism—are thematically interlinked elements in both destabilizing the confident authority of a scientific worldview.

Mina defies official pronouncements about women's mental and neurological weakness not only in her astute secretarial work but also, simultaneously, in the fact that her brain resists the submission of conscious will entailed in automatism.[37] In an important essay focusing on Mina's clerical function, Jennifer Fleissner sees her as a model for the late Victorian secretary, a machine-like body allowing for the illusion of disembodied masculinity in the modern business office. In this account, secretarial writing does not threaten but is instead continuous with women's presumed true vocation—wife- and motherhood—because, when performed well, it is, or rather appears to be, neither work nor thinking, but just another exercise of domestic and social grace. In concluding that Mina as secretarial exemplum *"know[s]*

---

37. For another reading of automatism in the novel, exploring its challenges to ideas of the soul, see Anne Stiles, "Cerebral Automatism, the Brain, and the Soul in Bram Stoker's *Dracula*," *Journal of the History of the Neurosciences* 15 (2006): 131–52.

how not to know" and in comparing Mina's typing to her hypnotic trance, Fleissner valuably intimates the divided structure of automatism purportedly essential to the typist's work.[38]

Yet we must remark that although Mina becomes automatic at the behest of Dracula, as a secretary she never really does. If secretarial work means appearing to know how not to know, in other words, Mina dramatically defies this expectation in visibly knowing how to know—visibly not only to us, but also to her allies. Even when in a patently automatic state, the hypnotic trance induced by Van Helsing, she looks to Jonathan like she is trying to piece things together. Prompted by a question, her "answer came dreamily, but with intention; it was as though she were interpreting something. I have heard her use the same tone when reading her notes" (312). As if she had not fully subdued her conscious tendencies, Mina's trance is marked by traits, intentionality and interpretiveness, that she also manifests when going over her stenography. The latter kind of task is hardly unthinking for Mina; it is in scrutinizing her own secretarial work and trance together that she is able to pick up Dracula's lost trail: "I read in the typescript that in my trance I heard cows low and water swirling level with my ears and the creaking of wood. The Count in his box, then, was on a river in an open boat." After she shares her memorandum on these thoughts—which reads like an elaborate flow chart—Van Helsing exclaims, "Our dear Madam Mina is once more our teacher. Her eyes have seen where we were blinded" (353).[39] His applause renders problematic the notion that Mina's allies remain in control of what she knows and does not know, "keeping her within the circle of knowledge *at their own command*";[40] here it is clearly she who commands what they know.

Mina has in fact been the group's leader all along, in strategic acumen if not in recognized practice. She has led in realizing immediately that catching Dracula necessitates sharing knowledge, while her allies lag behind, only fully learning the lesson through painful trial and error. Asking Seward to let her transcribe his journal, she reasons, "We need have no secrets amongst

---

38. Jennifer Fleissner, "Dictation Anxiety: The Stenographer's Stake in *Dracula*," *Nineteenth-Century Contexts* 22 (2000): 446. Fleissner notes a likeness between the seemingly untainted secretary and what she calls the "innocent female body" of the machine-like, "civil-listening" telephone operator discussed by Martin in *"Hello, Central?"* (451 n. 15).

39. For excellent close readings of Mina's interpretive secretarial work, see Alison Case, "Tasting the Original Apple: Gender and the Struggle for Narrative Authority in *Dracula*," *Narrative* 1 (1993): 222–43; though I take issue with Case's conclusion that the novel progressively tamps down Mina's intellect and reaffirms gender norms.

40. Fleissner, "Dictation Anxiety," 443.

us; working together and with absolute trust, we can surely be stronger than if some of us were in the dark" (223). Incidentally, this is another instance in which we see the possibilities, for good or for ill, for publicity in the medium's networked position. Mina's typing, as in the case of Seward's journal, unifies the group by releasing to them sometimes quite personal documents, "making private experience publicly available," as Peter K. Garrett observes.[41] In addition, it is another demonstration of how much Mina's astuteness dovetails with her secretarial work. By contrast, the other group members at first underrate the importance of mutual information: their most costly mistake, the one that enables Dracula to attack her, is keeping her uninformed. When the men decide immediately after the attack to bring her back in the know, her declaration that "there must be no more concealment... Alas! we have had too much already" puts a fine point on Van Helsing's folly in having excluded her in the first place and in turn on the naturalness of his handing over the title of "teacher" to her (290).

That Dracula is able to control Mina's mind says more about his occult potency than it does about her cerebral strength. In fact, Dracula's automatizing of her is partially motivated by his recognition of her craftiness: "And so you, like the others, would play your brains against mine....You are to be punished for what you have done. You have aided in thwarting me; now you shall come to my call" (287–88). Mina's strong sympathetic affinity with her allies makes warping her will the Count's best revenge against them. But vampiric captivation in this novel is otherwise indifferent to gender: the men are just as vulnerable to it as she is. Jonathan strains in Dracula's castle not to succumb to the entrancement of the dancing dust—"I was becoming hypnotized!" (44)—from which the trio of vampire women threatens to reemerge. Van Helsing, too, feels their power when he prepares to stake them, imagining that other men at the same task would likewise "delay, and delay, and delay, till the mere beauty and the fascination of the wanton Un-Dead have hypnotize him" (369).

Lessons of the Communications War

Vampiric hypnosis and Van Helsing's share a formal similarity, yet they differ substantially in their efficacy. But as in the initial assessment of Mina's neural capacity, there is a scientific arrogance that prevents the group from

---

41. Peter K. Garrett, *Gothic Reflections: Narrative Force in Nineteenth-Century Fiction* (Ithaca: Cornell University Press, 2003), 128.

understanding this reality. Once he realizes that Van Helsing is using her as a reverse conduit, Dracula closes the channel to keep them from following his movements. Van Helsing quickly realizes what the Count has done but boasts that it will not work: "his child-mind only saw so far...He think, too, that as he cut himself off from knowing your mind, there can be no knowledge of him to you; there is where he fail!" (343). Once again Van Helsing leans on an evaluation of brain power, with Mina chiming in, "The Count is a criminal and of criminal type. Nordau and Lombroso would so classify him, and *qua* criminal he is of imperfectly formed mind" (342). The assured reference to criminal anthropologist Cesare Lombroso and his disciple Max Nordau implies their expert knowledge of the mind, and a faith in it to which even Mina cannot help but be in thrall, even as this kind of thing condemns her. Yet this sense of certainty fails the group by causing them to overestimate their control over the situation. For plot events just subsequent to Van Helsing's boast suggest that it is in fact he, not the Count, who has miscalculated: attempts to employ Mina hypnotically begin to yield only fitful results immediately after the Count cuts her off. The next one requires "a longer and more strenuous effort on the part of Van Helsing than has been usually necessary" (344); the following morning the "hypnotic stage was even longer in coming than before; and when it came the time remaining until full sunrise was so short than we began to despair" (345). By the third try, Seward is beginning to fear "that her power of reading the Count's sensations may die away, just when we want it most" and that it is her "imagination" that is really dictating the substance of her reports (345).

Mina's unresponsiveness may also be due to her progressive enslavement to Dracula as she nears his homeland. But by either account, in fact, the vampire has successfully combated his foes in a head-to-head conflict over her psyche, reminding us of the Orient's age-old mastery of unconscious subtleties. Conversely, the situation points up the deficiencies of Occidental science's attempts to annex this field of knowledge. Jonathan indicates that the group relies on Charcotian models of hypnosis when he notes about Van Helsing's first attempt to entrance Mina, "I felt that some crisis was at hand": for Charcot, the hypnotic state was marked by identifiable "crises" corresponding to hysterical crises or stages of disease.[42] Jonathan also notes that the professor's "forehead was covered with great beads of perspiration" (312), which suggests that, even before Dracula has tampered with the connection, the group's method is a laborious undertaking. There is a marked distinction between the Count's

---

42. Adam Crabtree, *From Mesmer to Freud: Magnetic Sleep and the Roots of Psychological Healing* (New Haven: Yale University Press, 1993), 165–68.

telepathically immediate psychical relays ("When my brain says 'Come!' to you, you shall cross land or sea to do my bidding" [288]) and Van Helsing's attempts, which require a litany of tedious questions to Mina ("Where are you now?...What do you see?...What do you hear?...What else do you hear?...What are you doing?" [312–13]).

The professor's questions amount to a clunky word processing that exactly mirrors his group's other communicative disadvantage: their dependence on cutting-edge technologies that, however, become inefficient, slow, or otherwise unreliable in a pinch.[43] Awaiting Dracula in Varna, they are stunned and demoralized to receive a telegram telling them, too late of course, that Dracula's getaway ship has been rerouted to Galatz—a change of plans ensuing from the vampire's expeditious discovery through Mina's psyche of their own plans (337–39). And for all their sophistication, their writing apparatus appear to give out on them, become unusable, in the end. Stationed at Varna, Seward laments, "How I miss my phonograph! To write diary with a pen is irksome to me" (335). Seward may miss his machine, but earlier he admits to a big problem with it, that it is fairly useless for long records of information because it has no mechanism for indexed retrieval. It says something about the foolhardy overconfidence of the British scientist that this only just hits him when he shows the device to Mina: "do you know that, although I have kept the diary for months past, it never once struck me how I was going to find any particular part of it in case I wanted to look it up?" (221). Now, in hot pursuit of the Count, Seward has to do without his phonograph presumably for a different reason, because it does not travel well. Mina's typewriter fares a bit better because it is a "'Traveller's' typewriter" (350); but as the group nears Dracula's castle, she is no longer writing in shorthand, much less typing, one imagines, probably because of the Count's increasing, counteracting control over her brain. Like Seward, Van Helsing, stepping in for her as event recorder, regrets having to resort to conventional handwriting: "as Madam Mina write not in her stenography, I must, in my cumbrous old fashion, that so each day of us may not go unrecorded" (363).

These shortcomings reinforce the superiority of the Count's communication methods and, at the same time, indicate the inadequacies of some of the most spectacular achievements of late Victorian civilization. Inasmuch as the group's media practices, hypnotic as well as technological, condition the way

---

43. Likewise, Mina's typescript can in another way be said to mimic her trance: it generates a cumulative text that, with its many alternating voices, suggests the logic of psychic dissociation—the unconscious production of what we now call multiple personalities—a phenomenon recognized by hypnotists and psychologists around the time Stoker was writing, as I explain in chap. 4. But here again, Mina defies customary renderings of unconscious channeling, as her dissociated text is the product of neither passivity nor pathology but rather an investigative drive.

they access, record, and process reality, the way they shape it into something called knowledge, their media failures symbolize and, more covertly, partially effect the failures of their systems of knowledge—their science—to apprehend and comprehend the actual.[44] That is, though it may be tempting to view *Dracula* as a narrative of the triumph of positive knowledge, in which the forces of modernity synthesize information to squeeze out of existence the mystery Dracula represents,[45] the triumph is illusory at best.

In his postscript to the novel, Jonathan contemplates the technological accumulation resulting from the battle: "there is hardly one authentic document! nothing but a mass of type-writing, except the later note-books of Mina and Seward and myself, and Van Helsing's memorandum" (378). The "mass of type-writing" has garnered much scholarly attention, but usually at the expense of what Jonathan tells us is appended to it. The typed mass connotes an overwriting of the vampire by state-of-the-art methods of information processing only if one ignores the non-integrated materials, for these comprise the very same documents the group was forced to produce when their more advanced writing technologies proved unfit for the task. The surplus documents hint at a world the products and tools of modern science only imperfectly assimilate or control, reflecting on the allies' flawed ways of interacting, learning, and knowing. Their misleading classificatory distinctions and their overweening faith in their hypnotic procedure jeopardize their cause by blinding them to what makes them strong (Mina's ability to systematize and to deduce) and to what makes them weak (their limited power over her unconscious). Their views are predicated on a seemingly unimpeachable principle of empiricism, for instance on the notion of the physical evidence of women's cerebral and nervous systems. Yet implicitly their method draws on cultural preoccupations that already hinder perception, such as Van Helsing's image of Mina's future maternity: "she is a young woman and not so long married; there may be other things to think of some time, if not now" (235).[46]

---

44. The helpful premise that media enable discursive perspectives grounds Friedrich Kittler's concept of the "discourse network"; *Discourse Networks, 1800/1900,* trans. Michael Metteer, with Chris Cullens (Stanford: Stanford University Press, 1990). For Kittler's own reading of *Dracula,* according to which the Count confronts the discursive intertwining of psychoanalysis, psychophysics, and typing, see "Dracula's Legacy," in *Literature, Media, Information Systems: Essays,* ed. John Johnston, 50–84 (Amsterdam: G+B Arts International, 1997).

45. For readings to this effect, see David Seed, "The Narrative Method of *Dracula,*" *Nineteenth-Century Fiction* 40 (1985): 61–75; Rosemary Jann, "Saved by Science? The Mixed Messages of Stoker's *Dracula,*" *Texas Studies in Language and Literature* 31 (1989): 273–87; and Thomas Richards, *The Imperial Archive: Knowledge and the Fantasy of Empire* (London: Verso, 1993), 57–65.

46. Greenway emphasizes the limitations in *Dracula* of science and the social views it buttresses in arguing for the novel's depiction of a paradigm shift away from Seward's narrow-minded "normal

Examining *Dracula*'s communications conflict leads us to see how inseparable the issue of Mina's gender status is from the novel's larger views on knowledge contingencies, ethnic identity, and cultural hegemony. Although the narrative is hardly a fairy tale of complete illumination, by the end the characters have gained an inkling of their epistemological constraints. Indeed, their potential for uneasiness is hinted as soon as Van Helsing calls Mina's brain a "man's brain." The insulting quality of the phrase should not prevent us from recognizing that it affronts scientific categories by distending them. The idea that a woman's body could house such a brain entails such a contortion as to effectively suggest, within the world of the novel itself, the terminology's insufficiency and concomitantly the insufficiency of the social and moral truths it proclaims. But it is again Jonathan's concluding note that especially intimates the group's new awareness. Here he corrects the confidence of the novel's head note that their records make seeming impossibilities "stand forth as simple fact": pointing to the typescript, he declares, "We could hardly ask anyone, even did we wish to, to accept these as proofs of so wild a story" (378). The statement is remarkable for its ambiguity. Is he excusing the reader for not believing this "wild story" because it contradicts what science says is possible? Or is he, rather, indicting the structure of scientific knowledge itself, the always necessarily questionable "proofs" presumed essential to belief and knowledge? The exclamation by Van Helsing that follows retains the ambiguity but seems in its almost petulantly defiant tone to push toward the latter meaning: "We want no proofs; we ask none to believe us!"

Then, in a declaration that at first seems a non sequitur but actually extends his dubious take on the fruits of empiricism, Van Helsing suddenly turns our attention to Mina, framing her relationship to her son in a strikingly chivalric way: "This boy will some day know what a brave and gallant woman his mother is" (378). Just paragraphs above Van Helsing has called Quincey Morris, that paradigm of masculinity, a "gallant gentleman." The echo of the epithet here frustrates attempts to view Mina as thoroughly refeminized by maternity. The point—especially given Van Helsing's next and final statement, that Mina's son already "knows her sweetness and loving care; later on he will understand how some men so loved her, that they did dare much for her sake"—seems not to make us choose between reading

---

science" toward Van Helsing's psychical research and similarly maverick theories. He notes for instance Van Helsing's willingness to grant the legitimacy of telepathy, claimed by some psychical researchers to be only a different aspect of hypnosis, which had itself risen from pseudoscientific status ("Seward's Folly," 222–26).

Mina either as a lady defended by a legion of knights or as a knight herself, but rather to reinforce the taxonomical challenge of her character: her chimerical combination (in Van Helsing's earlier words) of a "man's brain" with a "woman's heart" (234). To the extent that Jonathan does mean to offer the narrative documents as "proofs," perhaps, then, the "wild story" they force the reader to consider is the story of not just the vampire, but along with him another supposed freak of nature, Mina's "man's brain." After all, it is this brain that has been principally responsible for devising the narrative itself— a narrative, moreover, that recounts her large part in finally tracking down and defeating the one male who ever managed to control her mind. Within the communications war, Dracula and Mina's "man's brain" inspire parallel paths from initial doubt to belief in the existence of things outside the realm of scientific possibility. The novel as a whole and the postscript in particular invite the reader to partake in this dual revelation.

Corelli's Heavenly Media

Dracula has well endured the transition from the Victorians to us, becoming a popular (if malleable) icon in media as dizzyingly diverse as those we see in the novel itself. The characters and works of Marie Corelli, by great contrast, have largely been forgotten—by scholarly readers as much as leisure ones, until the last decade—despite her stunning popularity during her own era with an audience ranging from the bourgeois consumer to eminences including William Gladstone and Queen Victoria. Of Corelli's thirty books, a good half were bestsellers; first editions often sold out as soon as they hit the shelves, and her debut novel, *A Romance of Two Worlds* (1886), went through fourteen editions in a decade.[47] Corelli was quite proud of her wide following and repeatedly contrasted it with her harsh critical reception. Her popularity especially draws me to her work because it hints at her engagement with the most culturally relevant narratives and images of her day. Significantly, Corelli offers some of the period's most sustained considerations of electricity as an element common to terrestrial and spiritual communications and, additionally, grants the female medium center stage, even while importantly rewriting the terms of that figure's mediation.

---

47. My data on Corelli's career and personal history come from Teresa Ransom, *The Mysterious Miss Marie Corelli: Queen of Victorian Bestsellers* (Phoenix Mill, UK: Sutton, 1999); Annette Federico, *Idol of Suburbia: Marie Corelli and Late-Victorian Literary Culture* (Charlottesville: University Press of Virginia, 2000); and Brian Masters, *Now Barrabas Was a Rotter: The Extraordinary Life of Marie Corelli* (London: Hamish Hamilton, 1978).

As a bestselling author, Corelli was keenly aware of media as agents of information transmission. Her career provides an early, potent example of a phenomenon we have come to associate with popular culture: the deliberate use of avenues of publicity for the construction and manipulation of celebrity identity. From her first encounters with the publishing world, Corelli propagated apparently false tales of her nationality and birth. A British woman who had grown up Minnie Mackay, she emerged before the public as half-Italian Marie Corelli and claimed that the man who was likely her illegitimate father was really her stepfather. As she aged, she consistently purported to be about ten years younger than she actually was. Anxieties about threats to her privacy and about the male press's stereotype of the homely female author meant that she habitually dodged camera-wielding paparazzi. When she did finally release a photograph of herself, she had had it retouched to smooth out her wrinkles and the curves of her body and to cast a brilliance on her hair, and alongside it ran her pledge that "no portraits resembling me in any way are published anywhere...invented sketches purporting to pass as true likenesses of me, are merely attempts to obtain money from the public on false pretenses."[48] No doubt Corelli was motivated by personal and professional pride here, but importantly, her feint also harmonizes with her novelistic philosophies regarding women's relation to media. By doctoring her image, Corelli decisively intervened in the visual story about her, reshaping the camera's "message," and this exemplifies her desire as an author to advocate women's self-present relationship to processes of mediated meaning production. Her writings portray women who, as media themselves, are especially presumed to have little control over what gets transmitted, yet these works have a didactic component in tacitly suggesting that women must have a personal and intellectual stake in the messages they convey.

In *The Soul of Lilith* (1892), the medium endures a Gothic physical entrapment in the London home of a Middle Eastern man who has become the thrill of the town due to his extraordinary scientific-cum-psychical abilities.[49] Lilith is hidden in an exotically furnished room; at first only a deaf Arab servant named Zaroba who serves as guard knows about the existence

---

48. Quoted in Federico, *Idol of Suburbia,* 43. Federico's first chapter reflects on Corelli's publicity-savvy self-construction and concentrates on her reaction to photojournalism. Reading the touch-up as simultaneously withholding and revealing, Federico concludes that it conveys the values informing Corelli's preference for the romance: her privileging of ideal illustrations over realism as avenues to the truth.

49. Elaine M. Hartnell discusses other ways in which Corelli's works interfuse Gothic stylings with her unique take on Christianity; see "Morals and Metaphysics: Marie Corelli, Religion and the Gothic," *Women's Writing* 13 (2006): 284–303.

of this girl whom the brilliant El-Râmi Zarânos found in a Syrian desert and snatched from death, and now keeps as a channel to the spiritual world. Lilith's body lies dormant while her soul wanders through the heavens that are its home. El-Râmi's occult science allows him to retrieve this soul at will, so that he can converse with it, "receive through the [body's] medium the messages of the Spirit," and thereby plumb infinite truths.[50] Eventually his brother, Féraz, learns of Lilith's chamber, but he is already generally privy to El-Râmi's ability for psychical control, which he thinks of as a run-of-the-mill Oriental talent: "as there were others of his race who could do the same thing, and as that sort of mild hypnotism was largely practised in the East, where he was born, he attached no special importance to it" (60). But Féraz is less aware of his own unconscious subjection to El-Râmi, which Corelli stresses through Féraz's effeminacy—his "femininely lovely" eyes and "womanish" gentleness (20, 57).

In this novel the ethnic incursion of the Orient is especially menacing in being localized onto a body, Lilith's, meant to represent core facets of Occidental culture—Christianity and whiteness—per se. Infiltration here comes down to the implied menace of sexual profanation. Six years after he brought her back to life, El-Râmi now fixates on her maturity into womanhood and fights back "wild desires" (228). Interestingly, Lilith may come from the same geographical region as he, but Corelli's disparate representations of her Middle Eastern characters frame his threat as one of miscegenation. On the one hand, the novel focuses obsessively on his dark skin and jetty eyes. On the other hand, Lilith is hyperbolically bleached out; we repeatedly return to her white throat, white hands, blue eyes, and blonde hair: in short, to her "white beauty" (351). When her soul climactically emerges freed from her body, El-Râmi is able, somehow, to make out "white outstretched arms," "radiant blue" eyes, and a "glittering flash of gold that went rippling and ever rippling backward, like the flowing fall of lovely hair" (362). The traces of whiteness with which Corelli endows some of her Middle Eastern characters appear to be honorary markers, indexing their embrace of Christianity. Féraz, a Christian by intuition, resembles El-Râmi but is "fairer-skinned" (20). Likewise, in Corelli's first book, *A Romance of Two Worlds,* the eyes of the Chaldean mystical physician Heliobas, whose theories illuminate Christ, are "a singularly clear and penetrating blue," and his sister, a follower of his creed, has skin that is "transparently clear—most purely white, most delicately rosy" (30, 99). In contrast are the tritely dusky features of unbelieving characters

---

50. *The Soul of Lilith* (1892; London: Methuen, 1905), 37. All further page references to *The Soul of Lilith* are parenthetical within the text.

like El-Râmi, who stubbornly resists Lilith's revelations of immortality and of Christ's love, and the "half-savage" Zaroba, who worships a vengeful pantheon (*Soul,* 387).[51]

El-Râmi's faithless materialism expresses itself both in his erotic fascination with Lilith's body and in his treatment of her as a scientific "experiment" (37), for which her body has been reduced to a precise instrument. A comatose "machine" sustained by "electric fluid" (37, 162), Lilith pushes to the very limit the figure of the automatic woman. Corelli has already represented women's spiritual messaging in *A Romance of Two Worlds,* though with several important differences, among them that Heliobas possesses the faith in the messages that El-Râmi lacks, generally marking him as an antithesis of the Oriental villain, and that the story is narrated from the medium's perspective, so that we are privy to her personal reflections on her role. The unnamed narrator is an improvisational pianist who, crippled by nervous infirmity, seeks help from Heliobas, whose therapy is based in electrical remedies gleaned from his close study of ancient Eastern thinkers. Soon the narrator becomes his spiritual disciple and absorbs the tenets of his elaborate cosmology, according to which God is an ever-productive source of electricity and each soul a spark emanated from this wellspring in the heavens. Many years after Genesis, his theory goes, human corruption and apathy prompted God to restore electric lines of communication with His creatures; hence Christ was sent to Earth to establish a connection similar to that made possible by submarine telegraphy: "this Earth and God's World were like America and Europe before the Atlantic Cable was laid. Now the messages of goodwill flash under the waves, heedless of the storms. So also God's Cable is laid between us and His Heaven in the person of Christ" (241). *Romance's* abundant technological metaphors for Christian worship—prayers imagined as telephone calls and the like (223)—culminate in this "Electric Principle of Christianity," which reinterprets Christ as, in effect, a Word dispatched telegraphically (235).

Corelli's introduction decries animal magnetism and modern spiritualism as spiritually bankrupt; nonetheless, *Romance* more or less directly borrows imagery from both. Heliobas possesses a "magnetic gaze" and a superior

---

51. The moral difference between Christian and non-Christian Middle Easterners is reinforced by Corelli's terminology. El-Râmi is frequently called an Oriental; by contrast, for Heliobas, Corelli generally opts for a term, Chaldean, explicitly linking him to the revered origins of Christianity. To the narrator's query, "You are a Chaldean?" Heliobas responds, "Exactly so. I am descended directly from one of those 'wise men of the East'... who, being wide awake, happened to notice the birthstar of Christ on the horizon before the rest of the world's inhabitants had so much as rubbed their sleepy eyes" (*Romance,* 67–68).

will that allows him to transfer, like Franz Anton Mesmer, electric vitality from himself to another (32). (Even the place in which Heliobas has made his practice, Paris, recalls mesmerism's namesake, who set up shop there for several years.)[52] Many of Corelli's novels absorb contemporary neurological theories, including ideas about nervous electrical energy.[53] These give her an avenue for representing the neural body as an information system, much as did séance-goers. One character in *Romance* imagines the narrator's sickness of the nerves as a failing in a "network" consisting of "electric wires on which run the messages of thought, impulse, affection, emotion" (52), a vision also suggesting the Victorian conceptual collapse of bodily and telegraphic networks. It is in part Heliobas's electrical regulation of this network—of what the narrator calls the "over-sensitiveness, the fatal delicacy, the highly-strung nervousness of the feminine nature" (84)—that eventually enables her to become a communication conduit, like the telegraphic Christ, from a higher world. Just before sending her on an out-of-body journey to the Great Circle of the universe, where God lives, Heliobas requests that she act as a heavenly messenger. Vaguely aware of some danger to his spiritual progress, he wants to take advantage of her journey to discover its exact nature: "you, when you are lifted up high enough to behold these things, may ... attain to the knowledge of it and explain it to me, when you return" (178). With the help of an angel named Azùl, she succeeds in the mission, telling Heliobas afterward, "Azùl said that I might deliver you this message: When death lies like a gift in your hand, withhold it, and remember her" (208). Though hazy at the time, these words, reiterated later, help to divert Heliobas from a morally erroneous path.

During her tutelage, the narrator learns that she has in fact long been a spiritual medium. Heliobas tells her that the spirits are the prime movers behind great art, whispering "divine messages—messages as brief as telegrams" (198); her own piano-playing is channeled from a spirit friend: "a flood of music seemed to rush to my brain and thence to my fingers, and I played, hardly knowing what I played" (229). Heliobas instructs her to yield completely to these higher forces: "if you once admit a thought of Self to enter your brain, those aërial sounds will be silenced instantly" (328). But though she appreciates the spirits' direction, the idea that she has no personal stock in her creation is a hard pill to swallow: "'It is lovely,' I said wistfully, 'all that music; but it is not *mine*;' and tears of regret filled my eyes. 'Oh, if it were only

---

52. Crabtree, *From Mesmer to Freud,* chap. 2.
53. Anne Stiles, "Nervous Electricity and the Neuron Doctrine in Marie Corelli's Fiction," unpublished manuscript.

mine—my very own composition!'" (140). The response prompts a gentle reproof from Heliobas—but Corelli has already implied that this generally enlightened man has a blind spot when it comes to valuing women's personal drive, in his earlier appraisal of them as "nothing but lumps of lymph and fatty matter" (85).[54] That insult stings the narrator enough for her to think of her exploration of the cosmos as a rejoinder, as she explains to Zara: "Your brother uttered some very cutting remarks on the general inaptitude of the female sex when I first made his acquaintance; so, for the honour of the thing, I must follow the path I have begun to tread. A plunge into the unseen world is surely a bold step for a woman, and I am determined to take it courageously" (146). Hence what might have otherwise been a self-abnegating act, her spiritual elevation, becomes just the opposite: an opportunity for personal and gendered self-declaration. In that way, too, she regains the sense of agency she so missed in imagining her music as a spiritual product and herself merely as its conduit.

Owning the Message

All this may seem an ill fit with the portrait of the titular character in *The Soul of Lilith,* in that that novel seems invested not at all in the possibility of Lilith's own perspective on what she channels and to glorify, rather, her pure mediation of loving Christian precepts. Her elevating powers of interceding between this world and the next could not be stated more plainly than when Féraz, returning home sickened by his experience of London debauchery one night, gratefully reflects on her sanctifying presence within his brother's house: "surely some Angel dwelt within and hallowed it with safety and pure blessing" (242). Looking at the fin de siècle, one is hard-pressed to find a writer more steadfast than Marie Corelli about the distinction between a sinful world at large and the home as a site of feminine virtue. In an essay in her *Free Opinions Freely Expressed on Certain Phases of Modern Social Life and Conduct,* one mourning the "decay of home life in England," Corelli might as well be channeling John Ruskin:

> A happy home is the best and surest safeguard against all evil... With women, and women only, this happiness in the home must find its

---

54. Heliobas's denseness on gender suggests surprising affinities with El-Râmi's objectifying perspective. Indeed, Alisha Siebers reads elements of claustrophobia and the Gothic in Heliobas's magnetic control. "Marie Corelli's Magnetic Revitalizing Power," in *Victorian Literary Mesmerism,* ed. Martin Willis and Catherine Wynne, 183–202 (New York: Rodopi, 2006).

foundation. They only are responsible; for no matter how wild and erring a man may be, if he can always rely on finding somewhere in the world a peaceful, well-ordered, and *undishonoured* home, he will feel the saving grace of it sooner or later, and turn to it as the one bright beacon in a darkening wilderness.[55]

Then again, where Lilith is concerned, the supposed profit to both sexes of "separate spheres" appears seriously questionable. Her domestic enclosure, because it really amounts to a Gothic incarceration by a sensual, forceful man, does not defend her femininity from moral perils but instead makes her more vulnerable to them. Further, her situation gives her very little opportunity for spiritual nurturance. Lilith is distinct from Mina Harker or Olga Trevelyan in seeming incapable of ever sharing in her manipulator's iniquity, yet his control nonetheless corrupts her sympathies in a different way: by restricting their usefulness. Locked up as she is, her holy messages fall on deaf ears— literally in the case of her attendant Zaroba, figuratively in that of El-Râmi, who stoutly denies the truths Lilith imparts. Thus she suffers from a physical as well as moral confinement, one that exemplifies the derangement of affective potential characterizing the medium's role in the late Victorian Gothic.

We should resist the impulse to simply label Lilith an angel in the house and then read her as the uncomplicated type of Corelli's ideals of femininity. To do so would be to disregard Corelli's fairly nuanced attitude toward women's roles as wives and mothers. Annette Federico uncovers a splitting in Corelli's writings on the subject: whereas some exalt female self-sacrifice for the family, others lash out against it as a cultural imperative guilty of fettering women and robbing them of their vigor.[56] In addition, even while affirming women's domestic duties, Corelli maintains that the idea of their mental limitation to these duties is one invidiously promulgated by men; as she writes in *Free Opinions:*

> [Woman's] place, says the didactic male, is the kitchen, the nursery, and beside the cradle. *Certes,* she can manage these three departments infinitely better than he can...But, as a matter of fact, there is hardly any

55. Marie Corelli, *Free Opinions Freely Expressed on Certain Phases of Modern Social Life and Conduct* (London: Constable, 1905), 220.

56. Federico, *Idol of Suburbia,* 117–20. For another analysis of Corelli's conflicting views on women and family, see Janet Galligani Casey, "Marie Corelli and Fin de Siècle Feminism," *English Literature in Transition, 1880–1920* 35 (1992): 162–78. Rita Felski also provides an intriguing perspective and offers helpful thoughts on *A Romance of Two Worlds* in particular, though I think she ultimately underestimates the significance of the heroine's spiritual journey as a response to anti-feminist ideology; see *The Gender of Modernity* (Cambridge, MA: Harvard University Press, 1995), chap. 5, esp. 135–36.

vocation in which she cannot, if she puts her mind to it, distinguish herself just as easily and successfully as he can if he will only kindly stand out of her way.[57]

She goes on to support women's assumption of a range of professions, including school inspector, surgeon, and lawyer.

Corelli's drive for professional female employment has much to do with her endorsement of women's intelligent activity and (as the title of her essay volume indicates) free self-expression. It is for just such tendencies that she admires the American woman: "She talks and laughs freely. She is not a mere well-dressed automaton like the greater majority of upper-class British dames."[58] Basically an automaton herself, Lilith seems pitifully remote from her author's conviction that women must strive to be "something greater than the mere vessels of man's desire" and to achieve "the very utmost extent of their moral and intellectual being."[59] If Lilith personifies the first of these goals, female morality, her complete submission to El-Râmi—her function, exactly, as vessel of his desire—mocks the second, female intellect. Even Lilith's name affronts the principle of women's self-determination, recalling as it inevitably does Adam's outspoken first wife, to whom Féraz at one point explicitly alludes (*Soul*, 337–38). Whereas that Lilith refused Adam's command to lie beneath him, protesting that she was his equal, El-Râmi's Lilith has no alternative in her mechanical state but to remain recumbent and receptive to her master's volition. Then there is the matter of this character's absence of a discrete, recognizable personality. El-Râmi's science has reduced her merely to a conduit of pure spirit. The novel is, after all, a tale about the *soul* of Lilith, not Lilith herself.

There is no denying that Corelli saw women as the cultural channels of morality as wives and mothers. Nonetheless, her depiction of the domestic angel highlights the drawbacks of the paradigm through caricature. Lilith's mental blankness suggests the hazards to personal development of women's socially mandated focus on family. Her sequestration points up their effectively captive positions within the home, which harms them as well as deprives many who might otherwise benefit from their influence.

In one apparent answer to these problems, Corelli's works carve out what we can identify as a compromise between the reactionary icon of the domestic angel and the progressive one of the female brain-worker: the inspired

57. Corelli, *Free Opinions*, 171.
58. Ibid., 118.
59. Ibid., 170–71.

woman artist. Federico remarks that many of Corelli's artists and writers are women and that this author believed in the feminine, ethical potential of her own profession as a means of moral elevation.[60] Corelli's depiction of female artistry in that sense widens women's responsibility as nurturers of goodness, beyond the narrow realm of the family and encompassing a larger audience of men and women. And for her this artist function is entirely compatible with mental determination. Consider *The Soul of Lilith*'s authoress character, Irene Vassilius. In one of his few moments of moral clarity, El-Râmi applauds Irene's instinct for transcendent truths: "Surely you must be a visionary, Madame!... and you see things not at all of this world!" (218). As a visionary she shares qualities with Lilith, emerging in her relations with El-Râmi as an "angel" and "messenger of peace" (430, 423). Yet where Lilith mindlessly obeys another, Irene is a satirist with a "strong will" and strikes others as "full of perfect self-possession" (195, 196). Hedonistic London socialites are put off by her independent thought and refusal to "accommodate herself to people's ways" (218), and the critics resist what she clearly evinces, women's potential for genius. "He and *his* Sex," she states of one of her carping male reviewers, "always consider the terrible possibilities to themselves of a badly-cooked dinner and a baby's screams. His notion about the limitation of a woman's sphere is man's notion generally" (194). Later she asks, "How is it that the 'superior' sex are cowardly enough to throw stones at those among the 'inferior,' who surpass their so-called lords and masters both in chastity and intellect?" (251).

The chaste and intellectual but critically undervalued female artist reappears in *The Sorrows of Satan* (1895) as the novelist Mavis Clare, the kind of woman with a "thinker's brain and an angel's soul."[61] Both Corelli's attraction to the character of the maligned authoress and her bristling at ideas of vacuous womanhood probably stemmed in part from the constant charges of dimwittedness leveled at her by her own reviewers. Here for example is one reviewer concurring with another: "A bold critic has thrown aside in disdain the novels of Miss Corelli, describing them as 'ignorant and illiterate.' We propose to make good this indictment."[62] One of her most common rebuttals was to accuse the press, a largely male group, of misogyny. In one essay she observes, "since I began my career six years ago, I have never had a word of

---

60. Federico, *Idol of Suburbia*, 99–101. On the culturally healing seer-artist as a recurring character in Corelli's fiction, see Richard L. Kowalczyk, "In Vanished Summertime: Marie Corelli and Popular Culture," *Journal of Popular Culture* 7 (1974): 854.

61. Marie Corelli, *The Sorrows of Satan* (1895; New York: A.L. Burt, 1923), 416.

62. Review of *A Romance of Two Worlds,* by Marie Corelli, and *The Christian,* by Hall Caine, *Quarterly Review* 188 (1898): 306.

open encouragement or kindness from any leading English critic," but later qualifies that her one "generous critic turned out to be—a woman—a literary woman too, fighting a hard fight herself."[63]

Implicitly Corelli meant to defy male views of what women writers can accomplish by typifying her own ideal of the inspired and inspiring, but also mentally self-possessed, female artist. Like her pianist-heroine carrying Azùl's message in *Romance*, Corelli is a transmitter of moral truths—at least she hints as much in her introduction to the novel: "if ever there was a time for a new apostle of Christ to arise and preach His grandly simple message again, that time is now" (xxii). (One reviewer, catching the gist of these lines, jeered at Corelli for affecting to "stand on a level with [Christ's] fishermen.")[64] When reading her works, with all their Christian plotlines, characters, and imagery, it is certainly hard not to surmise that their goal is as much to educate spiritually as to entertain. To a later edition of *Romance*, Corelli appended several fan letters thanking her for her uplifting tale and requesting more information about its electric creed. To her mind, obviously, these were sufficient evidence of the spiritual dearth in modern society and the necessity of her authorial mission:

> Scarcely a day passes without my receiving more of these earnest and often pathetic appeals for a little help, a little comfort, a little guidance, enough to make one's heart ache at the thought of so much doubt and desolation looming cloud-like over the troubled minds of many who would otherwise lead not only happy but noble and useful lives. When will the preachers learn to preach Christ simply—Christ without human dogmas or differences? (351)

It is hard to miss the audacity of Corelli's insinuation: that her writing fills a lack created by incompetent male clergy. She hardly concealed her contempt for this group, whom she found factious, selfish, and even secretly vicious (as in the case of the adulterous Reverend Anstruther in *The Soul of Lilith*). In calling for a corrective to "ministers who only minister to themselves and their own convenience,"[65] she prescribes for the services of the moral woman author, who can succeed where male clerics have failed in making divine wisdom available to humans. (Some clergymen apparently took her up on

---

63. Marie Corelli, "My First Book," *Idler* 4 (1893): 243. On Corelli's perception of her underestimation by the male press, see also Federico, *Idol of Suburbia*, 19.

64. Review of *A Romance of Two Worlds*, 306.

65. Corelli, *Free Opinions*, 344.

her offer of spiritual aid, excerpting and discussing her works during their sermons.)[66]

Corelli's particular framing of her authorial mission requires her to retool the standard image of the mediating woman to reserve for her a personal intelligence normally denied as both unfeminine and inimical to the communication process. In a footnote to an essay on her writing of *A Romance of Two Worlds,* she details the process by which she composes her novels:

> Since writing the above I have been asked to state whether, in my arrangements for publishing, I employ a "literary agent" or use a "typewriter." I do not. With regard to the first part of the query, I consider that authors, like other people, should learn how to manage their own affairs themselves, and that when they take a paid agent into their confidence, they make open confession of their business incapacity, and voluntarily elect to remain in foolish ignorance of the practical part of their profession. Secondly, I dislike type-writing, and prefer to make my own MS distinctly legible. It takes no more time to write clearly than in spidery hieroglyphics, and a slovenly scribble is no proof of cleverness, but rather of carelessness and a tendency to "scamp" work.[67]

This passage shows Corelli putting herself in the feminized position of communication medium, yet also revising that position to incorporate an element of rational autonomy. Note, first, that she rebuffs the picture of herself before a typewriter. Further, she is adamant about her own involvement in her business dealings. Through both these gestures, Corelli detaches herself from an association with the clichéd woman typist, who is presumed to do her work mechanically: to be employed in the thick of the business world but with little real understanding of the proceedings around her. Yet Corelli does not forsake the figure of the textual medium altogether. Quite the opposite, she makes a point to conjure a picture of herself as scribe, meticulously recopying her work in order to produce a "distinctly legible" manuscript. In fact, an entire page of the article features a sample of her tidy script of a page of *Romance,* with the caption "Fac-simile of Marie Corelli's MS. as prepared for the press."

Elsewhere in the essay, she elaborates on her painstaking clerical efforts:

> I generally scribble off the first rough draft of a story very rapidly in pencil—then I copy it out in pen and ink, chapter by chapter, with

---

66. Ransom, *Mysterious Miss Marie Corelli,* 75, 84–85.
67. Corelli, "My First Book," 252.

fastidious care...I find, too, that in the gradual process of copying by hand, the original draft, like a painter's first sketch, gets improved and enlarged. No one sees my manuscript before it goes to press...I correct my proofs myself.[68]

For Corelli, there is a concurrence between the work of transcription and her work as an author: her copying by hand gives her an occasion to improve and enlarge upon her original ideas. By insisting in this way on her role as her own secretary but then imbuing that role with creative activity, she divorces the feminine station of communication medium from familiar portraits of typing automata, transforming it into one of mental acuity and power.

Corelli serves as her own middlewoman, exerting unmitigated presence in the process by which her story reaches its audience ("No one sees my manuscript before it goes to press"), and this is also, I think, a way for her to underscore her notion of her art.[69] The aim is not just to invest feminine mediation with the intelligent qualities of authorship, but also to invest authorship with the affectively proximate qualities of feminine mediation. "I have always tried to write straight from my own heart to the hearts of others," she writes in the same essay; "I have come, by happy chance, straight into close and sympathetic union with my public."[70]

Her desire to speak "straight" from herself to them goes some way toward explaining her odd self-conflation with her narrator in certain sections of *Romance*. When in the prologue we hear that what will follow is an autobiography—"a plain history of strange occurrences happening to one's self" (2)—the speaker seems to be the narrator-heroine. But then the same speaker declines only two pages later to "hold myself answerable for the opinions expressed by any of my characters," as if perhaps she were Corelli herself and the book merely a fiction (4). Contributing to this confusion, in the appendix Corelli added to a later edition, she seems to assume her narrator's lessons with Heliobas as her own experiences, or at least akin to them, when she announces her reluctance to "set myself forward in any way as an exponent of high doctrine in which I am as yet but a beginner and student" (339). The artistic, half-Italian narrator of *Romance* already doubles the legendary

---

68. Ibid., 248.

69. Ransom's biography occasionally indicates that Corelli did at least eventually employ a secretary who among other things transcribed her writings (just as she eventually employed a literary agent). But this should not distract from, and indeed probably only clarifies, Corelli's objective in these statements early in her career about preparing her own manuscript: to create for her public a certain image of herself best suited to her gender and authorial ideals.

70. Corelli, "My First Book," 239, 250.

Marie Corelli (who had at one time tried her hand at improvisational piano), and the narrator's namelessness makes it even easier for Corelli's audience to imagine that the author herself is the "I" about whose adventures they are learning. By framing the story in a voice that cultivates this identification, Corelli creates the illusion of talking immediately to her readers, thus working toward the intimacy with them in which she would later express such pride. At the same time, she casts her narrator as a stand-in for herself as a channel of moral messages. Narrator again fuses with author when the former reaches out at the end of her tale to deliver a "friendly warning" to her spiritually delinquent modern readers (335).

Corelli's narrator linguistically mediates her novel, and indeed, every narrator is essentially a medium, conveying a message-text from an author-sender to a reader-receiver. The narrator's spiritual mediation for Heliobas in *Romance*—like Mina Harker's typing for her allies, which makes up and relays the narrative of *Dracula*—simply draws attention to this fact by allegorizing it. Importantly, Corelli collapses herself with her narrator to take advantage of the greater closeness to the reader that the mediating position entails; in doing so, she alerts us to the potential for creating sympathy within the mediating narrative voice.[71] Like George Du Maurier, whose work I turn to in the next chapter, Corelli was a popular author who understood the pivotal affective role of media, and specifically the mediating narrative voice, in the authorial relationship with the public. Without detracting from her sincere concern for her readers' virtue, we may acknowledge that *Romance* shows her exploiting narration, much as she did photography, as a medium for creative self-disguise: for putting forward an exemplary version of herself. The closeness she claimed with her readers should not obscure her actual careful policing of the divide between the public and her privacy. Striking a balance between the public and the private was crucial not only to her as an individual but also to her concept of female artistry, as a career that recasts as a professional endeavor women's traditionally familial vocation of moral caretaking.

Focusing on mediation often takes us back to issues of privacy and publicity, the hidden and the revealed, because media always potentially function

---

71. In this respect, and in using narration as an instrument for Christianizing her readership, Corelli offers an unusual spin on what Robyn R. Warhol takes to be a nineteenth-century feminine mode of narration. Warhol argues that because female authors were deprived of other avenues of social efficaciousness, they found an outlet in an "engaging" first-person narrative voice. Imitating evangelical persuasiveness, Elizabeth Gaskell and others sought through narration to nurture sympathy for their characters and stories, and so to influence culture in the same way they were expected to exercise influence domestically. See *Gendered Interventions: Narrative Discourse in the Victorian Novel* (New Brunswick, NJ: Rutgers University Press, 1989).

as channels for accessing what is too off-limits or remote to lay a hold of otherwise. By a similar logic, the mediating woman lends herself to tales of exotic trespass onto white culture—its land, its people, its sacred beliefs. El-Râmi's profane, intrusive use of Lilith to probe Christian mysteries is a less concentrated version of Dracula's attempt to demonize English hearts through Mina Harker. Perhaps it is the especially intense form of female automatism in these works that prompts their confrontation of assumptions of women's limited self-presence. Like Mina's combination of "man's brain" and "woman's heart," the "thinker's brain and angel's soul" of *The Sorrow of Satan*'s Mavis Clare reinforce the idea of feminine moral sentiment common to Gothic tales of foreign entrancement even while correctively supplementing portraits of the helpless psychical thoroughfare.

❧ CHAPTER 3

Du Maurier's Media

The Phonographic Unconscious on the Cusp of the Future

> That Trilby was just a singing-machine—an organ to
> play upon—an instrument of music—a Stradivarius—
> a flexible flageolet of flesh and blood—a voice, and
> nothing more—just the unconscious voice that
> Svengali sang with—for it takes two to sing like
> La Svengali, monsieur—the one who has got the
> voice, and the one who knows what to do with it.
>
> —George Du Maurier, *Trilby*

By 1898, *Dracula,* with its vision of a magical
man who enters onto unwelcoming soil through an automatized woman,
is reworking the basic outlines of a story already made extremely popular
in George Du Maurier's *Trilby* (1894). Svengali seems to hail from Austria,
yet he is principally ethnically marked (and maligned) as Jewish; his origin
is the "mysterious East."[1] His mesmeric powers allow him to transform the
ingénue Trilby into the world-famous La Svengali, who produces harmonies
whose intricacies he grasps implicitly but is incapable of vocalizing himself.
With her as his instrument, he realizes his fantasy of overpowering even the
most exalted figures of Western society. "Svengali will go to London *himself.*
Ha! ha!" he gloats early on, sounding something like Dracula. "Hundreds of
beautiful Engländerinnen will see and hear and go mad with love for him—
Prinzessen, Comtessen, Serene English Altessen. They will soon lose their
Serenity and Highness when they hear Svengali!" (67). His sudden death
turns Trilby's London performance into a fiasco, but before then he succeeds

---

1. Du Maurier, *Trilby,* ed. Daniel Pick (1894; New York: Penguin, 1994), 259. All further page
references to *Trilby* are parenthetical within the text. Like Dracula, Svengali may be a fearful rendering of the many Eastern European Jews immigrating into England at this time; see Jules Zanger, "A
Sympathetic Vibration: Dracula and the Jews," *English Literature in Transition, 1880–1920* 34 (1991):
32–44, and Daniel Pick, *Svengali's Web: The Alien Enchanter in Modern Culture* (New Haven: Yale
University Press, 2000), chap. 9.

in enrapturing many a "personage" in Paris and elsewhere, compelling their worship through her voice (188).

Svengali's accomplice, Gecko, cannot pinpoint the "singing-machine" that Trilby's "unconscious voice" reminds him of. As the recorder of Svengali's knowledge, which then plays back that knowledge acoustically, her unconscious resembles none of the apparatus Gecko runs through so much as it does the phonograph.[2] That Du Maurier should think of this machine as female may come as little surprise given the Victorians' tendency to think of females—typists, operators, séance mediums—as machines. But the gendering of the phonograph in *Trilby* also has something to do with the operation it allegorizes, hypnotism. In an 1890 essay in the *Quarterly Review* titled "Mesmerism and Hypnotism," the author observes, "By placing one hand on the forehead and the other on the neck, the patient is transformed into a 'veritable phonograph,' giving back with perfect exactness every sound uttered by the [hypnotist]."[3] Both hypnotism and the phonograph deal in the recording and reproduction of information; the unconscious mind, tapped by the hypnotist, doubles the phonograph's retentive wax cylinder. Popular as well as medical lore identified women as the better subjects for this process, for about the same reasons as held true in spiritualism: women's inherent sensitivity and relative lack of personal agency helped them to become unconscious easily and to succumb to the suggestions of the usually male hypnotist, who was often described as imposing his own will on another. Because late Victorians saw women as ideal hypnotic subjects, and hypnotism on the model of the phonograph, it was probably only a short leap to imagining the phonograph as a woman.

French author Villiers de l'Isle Adam's novel *L'Eve future* (1886) also attests to the interchangeability in the fin de siècle imagination of female phonographs and hypnotic subjects. In this novel a fictional Thomas Edison has provided his android construction Hadaly with an electrical network of "nerves" and "veins," along with two "lungs" that are really two phonographs. Edison is visited at Menlo Park by an old friend, Lord Ewald, who has fallen in love with an exquisitely beautiful singer with an achingly mediocre personality.

---

2. For John M. Picker, Trilby's resemblance to the technology connotes its association with eroticized womanhood, as does the phonographic android in Villiers de l'Isle Adam's *L'Eve future;* see *Victorian Soundscapes* (Oxford: Oxford University Press, 2003). Lisa Gitelman also notes Trilby's phonographic demeanor; she implies that Du Maurier is a kind of automatic writer whose mechanistic production of pulp melodrama, like Trilby's performance, confounds attempts to pinpoint authorial agency; *Scripts, Grooves, and Writing Machines: Representing Technology in the Edison Era* (Stanford: Stanford University Press, 1999), 212–13.

3. "Mesmerism and Hypnotism," *Quarterly Review* 171 (1890): 249–50.

With his friend despondent about this clash between Alicia's body and soul, Edison sees a chance to realize Hadaly's potential: he will reconstruct her as the singer's clone; the then gorgeous android will possess a fittingly gorgeous personality, as composed of her splendid phonographic recordings. Edison has selected these, the words of brilliants artists and philosophers, according to his own understanding of the Ideal (the supposed translation of her name), which is to be compounded with the unique tastes of Lord Ewald. As a faux singer who looks animated but really functions automatically, vacantly repeating the sensibilities of a male creator, Hadaly strikingly anticipates La Svengali. But the similarity runs still deeper, for Villiers clearly presents the android as a version of the hypnotic subject. Edison shows an aptitude for hypnosis, and at one point compares Hadaly when switched off to "those sleepwalkers who can be made to pose by their hypnotists as if in a cataleptic trance."[4] Given Du Maurier's French youth and his ongoing attraction to all things French (all three of his novels are partially set in France), and his frequent conversations about French literature with his good friend Henry James,[5] it is quite possible that Du Maurier knew of Villiers's novel. In any event, if Villiers's novel dramatizes recording technology as an approximation of hypnotism, Du Maurier's novel confirms this commutatively by rendering hypnotism as a kind of recording technology.

In literally delving into the unconscious mind, hypnotism raised numerous questions about our subterranean connections to other people, worlds, and times. This chapter examines repeated images of the recording unconscious in Du Maurier's novels as they enable him to explore these connections. I am centrally concerned with how the phonographic unconscious becomes a means for his fictions to look ahead to the future. In *Trilby*, this translates as an engagement with aspects of the modern. Du Maurier is writing on the threshold of a new century, and his contemplation of modernity through recently invented media is at bottom a contemplation of mediation itself. This author was obviously fascinated by burgeoning media: all three of his novels feature characters who become unconscious channels between distinct periods, places, or individuals, with some progressive communication method offered as the analogue of their accomplishment. Phonographic Trilby becomes a figure for probing new turn-of-the-century criteria for artistic production, especially the popularizing potential of mediation and its

---

4. Auguste, Comte de Villiers de l'Isle Adam, *Tomorrow's Eve*, trans. Robert Martin Adams (Urbana: University of Illinois Press, 1982), trans. of *L'Eve future*, 142–43. All further page references to *Tomorrow's Eve* are parenthetical within the text.

5. Leonée Ormond, *George Du Maurier* (Pittsburgh: University of Pittsburgh Press, 1969), 400.

effects on capitalistic as well as aesthetic value. The novel can seem remarkably prophetic for the issues it raises about the role of popular media in the modern world. We are still replaying La Svengali's song and wondering how it resonates with us today.

Du Maurier's other two novels, *Peter Ibbetson* (1891) and *The Martian* (1897), imagine the future in a much grander sense. Here the phonographic unconscious connects to a spiritual beyond. Because the phonograph operates on a timeline, generating a record that promises perpetuation, Du Maurier's embedding it, as it were, within the individual foregrounds issues of our unconscious relation to time and history. Roger Luckhurst rightly points out that we have become too used to thinking about the Victorian unconscious in terms of degeneration and the primitive. The respected British psychical researcher Frederic Myers typified a counter-trend: for him the unconscious mind sends up to our consciousness fleeting hints—telepathy, hysteria, and other tokens of heightened sensitivity and perception—of our evolutionary advance, or "*progenerate*" capacities.[6] Du Maurier represents a cross between these two perspectives, but with the emphasis ultimately on amelioration. As *Peter Ibbetson* particularly stunningly implies, in the unconscious lie both vestiges of our racial past and the stuff on which our spiritual evolution is built. This is less a paradox than it might seem. The temporal concern of the phonograph is really bidirectional, as it equates remembrance, a record of the past, with a view to posterity, the future. In Du Maurier's novels, similarly, what we become is contingent on what we have been.

In *Peter Ibbetson* and *The Martian,* Du Maurier employs the phonographic unconscious to imply the fullness and transcendence of the self. For hypnotists and other investigators of the unconscious, likewise, the phonograph, along with the camera, served as a metaphorical means of envisioning mnemonically continuous selfhood and even, by extension, some form of spiritual continuity and connection to an immortal plane of being. These associations should inform our view of turn-of-the-century media and of how they shaped people's understanding of themselves and their communications. My argument later in the chapter suggests how different, and ontologically plentiful, what Friedrich Kittler terms the discourse network looks once we read it against the culturally significant backdrop of hypnotism, occultism, and psychical research.

---

6. Roger Luckhurst, *The Invention of Telepathy: 1870–1901* (Oxford: Oxford University Press, 2002), 110, 184.

Automatic Song and the Enamored Crowd

The automatic behavior at the heart of La Svengali's performance is central to *Trilby*'s view of art in a rapidly changing social and commercial environment (one contemporaneous with the novel's publication though not its action). Here again *L'Eve future* provides an instructive comparison. Villiers's Edison projects the production of multiple automata on the basic model or "substratum" of Hadaly—an entire slew of Stepford mistresses: "Only the first Android was difficult.... There's no doubt that within a few years substrata like this one will be fabricated by the thousands; the first manufacturer who picks up the idea will be able to establish a factory for the production of Ideals!" (147). Edison's dream of an android factory quivers with the industrial zeitgeist of the late nineteenth century. Hadaly's recording cylinders allow for the material duplication and *re*duplication of the Ideal, converting an abstraction into a tangible, mass-produced commodity.

An identical principle governs in *Trilby*. Once recorded, the phonographic woman's vocalization of her creator's influence is potentially limitless; hence it becomes the basis for any number of marketable reproductions. It is through these, Trilby's many packed shows, that Svengali acquires his pricey furs and the three thousand pounds he leaves unbequeathed at his death. Her song is what Walter Benjamin would have called an artwork susceptible to mechanical reproduction, an industrial-age artwork. Yet importantly, it succeeds by *looking* each time like an auracular creation. Her audience is touched to see her approach stage front and "put her hand to her heart quite simply and with a most winning natural gesture, an adorable *gaucherie*—like a graceful and unconscious school-girl, quite innocent of stage deportment" (190). Trilby behaves in a way that seems momentary, "natural," off-the-cuff, but Du Maurier's reference to the "unconscious school-girl" slyly gives the game away. Each of her recitals, though seemingly unique and ephemeral, has actually been carefully fashioned by Svengali to be, automatically, an equally stupendous production every time. The origin of Trilby's song looks obvious (Trilby herself), but it is really obscured and incidental, as Benjamin affirms to be the case for mechanically reproduced works of art generally. Moreover, to all appearances, Trilby basks in the ardor of her spectators, returning the warmth of their gaze "with her kind wide smile" (192), but in truth her gaze is sightless. The automaton cannot reciprocate another's regard and so cannot enter into a dynamic with her audience, a fact that—as with the film actor performing before a camera instead of a live audience—deprives her of the aura that embraces the stage performer: "To perceive the aura of an object we look at means to invest it with the ability to look at us in return."[7] Finally, Trilby's implication in the mechanisms of industrial art becomes most clear when her image, with all its apparent tenderness, is photographically reproduced and sold for the benefit of her devoted audience: "A crowd of people as usual, only bigger, is assembled in front of the windows of the Stereoscopic Company in Regent Street, gazing at presentments of Madame Svengali in all sizes and costumes. She is very beautiful—there is no doubt of that; and the expression of her face is sweet and kind and sad" (222).[8]

But my aim here is not to affirm Benjamin's theory of mechanically reproduced art by supporting it with more evidence. Rather, I want to suggest the historicity of such theories: Du Maurier's novel fictionalizes developing media in ways that can help us to recognize and analyze how their social milieu conditions suppositions about their impact. That *Trilby* shares features with Benjamin's work results from the fact that both are meditating on the effects of modernity, technology, and capitalism on art and artistic reception; absorbing dominant conceptions of gender; and figuratively relating the one to the other. As I will be arguing, *Trilby* forecasts later assessments—among other respects, in implying the feminine qualities not just of the popular (popular art and social formations) but of popularizing media themselves.

Incidentally, phonographic Trilby is not the only feminine medium in the novel. Her rabid audience is one as well. Their "frantic applause," "ecstatic delight," and increasingly "wilder enthusiasm" all signal a class leveling and a generalized hysteria (one renowned composer in the audience is explicitly described as quite "hysterical") (194–95). This is, once again, the mass incarnation of the automatic woman that shaped ideas of the crowd—of a social body devoid of personal rationality and governed by emotional response.

---

7. Walter Benjamin, "On Some Motifs in Baudelaire," in *Illuminations: Essays and Reflections*, ed. Hannah Arendt, trans. Harry Zohn (New York: Schocken, 1968), 188.

8. The high demand for these purchasable memorabilia of La Svengali suggests the "'spell of the personality,' the phony spell of a commodity" that Benjamin claims is cast around the mechanically reproduced performer (the film actor) in the absence of authentic aura; "The Work of Art in the Age of Mechanical Reproduction," in *Illuminations*, ed. Arendt, 231. As Mary Russo notes, Trilby's "highly profitable aura of celebrity" is partly enabled by the photographic "mass-reproduction" of images of her best physical asset, her foot; *The Female Grotesque: Risk, Excess, and Modernity* (New York: Routledge, 1995), 148. In another vein, it bears remarking that a dismemberment parallel to this figurative one of Trilby's foot occurs in *L'Eve future*, when Edison presents to Ewald an artificial female arm as proof of his android's technical perfection (Villiers, 59–60). In both Du Maurier's and Villiers's works, the amputation of the female limb, a prime locus of movement, connotes the transformation of the animated woman into a controllable, marketable expression of male subjective creation. On, similarly, Svengali's conversion of Trilby into a kind of statue, see Gail Marshall, *Actresses on the Victorian Stage: Feminine Performance and the Galatea Myth* (Cambridge: Cambridge University Press, 1998), 156–62.

On one level, this incarnation attests to Andreas Huyssen's claim that crowd theory exemplified a turn-of-the-century tendency to see mass culture as feminine (and degraded).[9] On another level, it betokens that the crowd was an instrument of communication, in which members shared states of mind, emotion, and sensation. Hence the quasi-telegraphic metaphors that crop up in descriptions continuing into the twentieth century. In his own writings, Benjamin clearly evokes a circuit running through the dense traffic of the urban center: "Moving through this traffic involves the individual in a series of shocks and collisions. At dangerous intersections, nervous impulses flow through him in rapid succession, like the energy from a battery. Baudelaire speaks of a man who plunges into the crowd as into a reservoir of electric energy."[10] Becoming part of the crowd means experiencing a feminine neuro-electrical sensitivity to one's surroundings.

Critics have often noted *Trilby's* intimation of the crowd and Svengali's representation of the mesmerically persuasive demagogue. But I would emphasize that Svengali's demagoguery is not direct: it absolutely requires the mediating presence of Trilby.[11] When he fantasizes early on about enthralling the highest strata of English society, he thinks he can achieve this goal by his own devices: "He will be all alone on a platform, and play as nobody else can play" (67). But as it turns out, Svengali will not be able to succeed alone on a platform. Because he is "absolutely without voice, beyond the harsh, hoarse, weak raven's croak he used to speak with" (38), and in order to ingratiate himself with people who snub him as a grimy "blackguard" (153), he must operate through Trilby, for reasons that have as much to do, as we will see, with her femininity as with his ethnic difference.

With the expansive reach of her song—which expands the scope of Svengali's self-publicity and cultural invasion—Trilby is effectively a

---

9. Andreas Huyssen, *After the Great Divide: Modernism, Mass Culture, Postmodernism* (Bloomington: Indiana University Press, 1986), 52–53.

10. Benjamin, "On Some Motifs," 175. As Tim Armstrong claims, moreover, such depictions hint at Benjamin's engagement with an attention-distraction binary that would become central to Modernist discussions of automatic writing. *Modernism, Technology and the Body: A Cultural Study* (Cambridge: Cambridge University Press, 1998), 216–19.

11. For examples of readings of the crowd in *Trilby*, see Alison Winter, *Mesmerized: Powers of Mind in Victorian Britain* (Chicago: University of Chicago Press, 1998), 339–41, and Pick, *Svengali's Web*, 71–75. Phyllis Weliver thought provokingly argues for Trilby's essential part in the effect over her audience in *Women Musicians in Victorian Fiction, 1860–1900: Representations of Music, Science and Gender in the Leisured Home* (Aldershot, UK: Ashgate, 2000), chap. 5, and also in her "Music, Crowd Control and the Female Performer in *Trilby*," in *The Idea of Music in Victorian Fiction,* ed. Sophie Fuller and Nicky Losseff, 57–80 (Aldershot, UK: Ashgate, 2004). Yet these pieces overstate Trilby's personal agency in her relationship with Svengali. Acknowledging her importance to Svengali's music-making need not interfere with recognizing her passive identity and instrumentality.

broadcast or mass medium, combining the power to record information with the power to disseminate it. As an instrument for amusement, the phonograph itself would thrive as a mass medium. But in its early years, when it could not only reproduce but also record, and when it exhibited a certain functional "plasticity"—its principal communicative modes not yet solidified through habituating social and institutional practices—the context for its usage was envisioned quite differently.[12] Edison and other designers mainly saw it as an alternative to the stenographer's shorthand—a Victorian synonym for which, *phonography,* sums up its historical link to the phonograph's method of inscribing sound (in the case of shorthand, the human voice).[13] As a trade journal editor observed, a particular value of using the phonograph as a dictating machine was that it would "becom[e] a confidential agent whenever secrecy may be required."[14] No need to worry about how mechanically the stenographer did her work when an actual machine could replace her. However, the market for office phonographs never took off. Instead the success of publicly installed coin-operated machines that replayed songs and other diversionary bits gradually revealed that people most appreciated the phonograph as a source of entertainment.[15] One noteworthy facet of Trilby's phonographic performance is that it combines the technology's originally intended use with its eventual one: Svengali's medium takes dictation and also reproduces it for others' recreation.

Seen in another light, it is not just music that Trilby communicates but the mesmeric state itself. By her enchanting song, she passes on her own fascination to her audience.[16] And through the audience's response, Du Maurier's novel theorizes, if obliquely, the natural affinity between mesmerism and capitalism.[17] In the first place, as the scene before the Stereoscopic

12. On the initial "plasticity" of modern sound technologies, see Jonathan Sterne, *The Audible Past: Cultural Origins of Sound Reproduction* (Durham: Duke University Press, 2003), 191–96.

13. On phonography/stenography as a means of sound recording, see Gitelman, *Scripts, Grooves,* chap. 1.

14. Quoted in Leonard De Graaf, "Thomas Edison and the Origins of the Entertainment Phonograph," *NARAS Journal* 8 (1997–98): 64.

15. On nickel-in-the-slot phonographs and their users, see Lisa Gitelman, *Always Already New: Media, History, and the Data of Culture* (Cambridge, MA: MIT Press, 2006), 44–56.

16. Athena Vrettos relates the audience's mesmeric reaction to late Victorian concepts of neuromimesis, a feminine condition in which one imitates the pathology of another. *Somatic Fictions: Imagining Illness in Victorian Culture* (Stanford: Stanford University Press, 1995), 103–5.

17. Other perspectives on the novel's conjunction of mesmerism, mass culture, and consumerism are in Pick, *Svengali's Web,* 83–91; Pamela Thurschwell, *Literature, Technology and Magical Thinking, 1880–1920* (Cambridge: Cambridge University Press, 2001), chap. 2; and Lara Karpenko, "Purchasing Largely: *Trilby* and the *Fin de Siècle* Reader," *Victorians Institute Journal* 34 (2006): 215–42.

Company in Regent Street implies, La Svengali—her mesmeric song and, metonymically, her self—is a product that can be bought and sold.[18] Her first onstage appearance to Little Billee and his friends is as a long-haired beauty "like those ladies in hairdressers' shops who sit with their backs to the plate-glass window to advertise the merits of some particular hair-wash" (189). From her redolence here and at the Stereoscopic Company of commodity culture, to Svengali's riches and the house conductor's promise to the audience just after her disastrous London performance that all "the money would be returned at the doors" (229), Du Maurier never lets us forget the commercial stakes of Trilby's performance. In the second place, the consumer appeal of that performance is inversely dramatized as a kind of mesmerism. When they encounter La Svengali, even her most lofty fans are prostrated and, as Sarah Gracombe aptly puts it, "feel spontaneously compelled to hand over their wealth," surrendering the very jewels off their necks.[19]

Even though we rarely articulate them as such, the conceptual links Trilby offers still shape our considerations of our buying habits. We are essentially debating just how mesmeric popular commercial practices are when we enter into lawsuits about the effects of tobacco advertisements and fast-food menus: exactly what degree of control does one have over his or her own will in the face of such allures? In a related manner, Trilby's broadcast performance prefigures and exaggerates the modern tendency to link mass communication to mesmeric domination.[20] Many marketers frankly capitalize on the notion that mass media promise an escapist loss of self. *Mesmerizing* and *hypnotic* are surely some of the most commonly used adjectives for promoting a new film or book.

---

18. Trilby thus bears a remarkable resemblance to a slightly earlier novelistic heroine, Verena Tarrant in Henry James's *Bostonians* (1886). Verena is billed by her father, who styles himself a mesmerist, as an inspirational speaker; as Richard Salmon has shown, moreover, different characters wrangle over her as a celebrity-commodity and vacant channel of seemingly intimate speech. *Henry James and the Culture of Publicity* (Cambridge: Cambridge University Press, 1997), chap. 1. In 1889, Du Maurier offered the plot of *Trilby* to Henry James before penning it himself (Ormond, *George Du Maurier*, 413), yet it is arguable that James had already written his own version of the story.

19. Sarah Gracombe, "Converting Trilby: Du Maurier on Englishness, Jewishness, and Culture," *Nineteenth-Century Literature* 58 (2003): 103. For Gracombe, this mesmeric effect derives from art, epitomized in the music of Jewish virtuosi.

20. On this linkage in modern thought, see also John Durham Peters, *Speaking into the Air: A History of the Idea of Communication* (Chicago: University of Chicago Press, 1999), 93–94. Jonathan Freedman calls *Trilby* a "remarkable adumbration of twentieth-century mass culture theory" without expanding on the point, focusing rather on how the novel models for a middlebrow audience their own reception of high culture as represented by the figure of the Jew; *The Temple of Culture: Assimilation and Anti-Semitism in Literary Anglo-America* (Oxford: Oxford University Press, 2000), 101.

Exploiting the Friendly Medium

What is particularly intriguing about *Trilby* and has yet to be examined is its implicit presentation of mediation as a crucial factor in creating popularity, in entrancing mass consumers. The medium, Marshall McLuhan famously declared, is never merely neutral. Similarly, it is the way art is transmitted that Du Maurier's work reflects as an up-and-coming and powerfully enthralling force. This is a novel intensely conscious of popularity and its mechanisms—hence perhaps its own extraordinary success first in the United States, where it was originally published, and then in England. Hence also the spate of products, adaptations, and parodies that came after it, inspiring what is becoming a critical commonplace about *Trilby,* that "the story reflects, disturbs, interacts with, its own overblown reception."[21] American psychologist Boris Sidis wrote of the public's fascination with the book in language that recapitulates its own sinister mesmeric events: "Who can tell of all the crazes and manias—such, for instance, as...the Trilby craze ...—that have taken possession of the American social self?"[22]

One significant way in which *Trilby* induced the love of the public had to do, as holds true within the novel itself, with the mediatory presence between artist-creator and audience. This achievement in mediation is legible in the context of turn-of-the-century developments in author-reader relations. As Barbara Hochman has argued, throughout most of the century, readers had essentially equated novels with the authors themselves, as if sitting down to read were like chatting with a friend, and authors had cultivated this perception through warm "I" narrators and direct addresses. But by the late 1800s, the rise of a large, diverse reading populace encouraged authors' retreat from the narrative and, by extension, from readers about whose identities and motives they had become more uncomfortable. Contributing to this sense of distance were new copyright laws that separated the text from its author as a possessible property, along with the increasing intermediary participation of editors and agents. Under all of these circumstances, Henry James and others began to formalize an impersonal, "objective" mode of realist narration in which characters and situations speak for themselves, and to rebuff as aesthetically

---

21. Pick, *Svengali's Web,* 36. For details of the *Trilby* "boom," see Emily Jenkins, "*Trilby:* Fads, Photographers, and 'Over-Perfect Feet,'" in *Book History,* ed. Ezra Greenspan and Jonathan Rose, vol. 1, 221–67 (University Park: Pennsylvania State University Press, 1998); L. Edward Purcell, "Trilby and Trilby-Mania, The Beginning of the Bestseller System," *Journal of Popular Culture* 11 (1977): 62–76; and *Trilbyana: The Rise and Progress of a Popular Novel* (New York: The Critic, 1895).

22. Boris Sidis, *The Psychology of Suggestion: A Research into the Subconscious Nature of Man and Society* (New York: Appleton, 1898), 364.

inferior fictions that suggest the author's hovering presence. At the same time, many readers were turned off by Jamesian diffidence and perpetuated a wide demand for close author-reader relations within the text. Consequently, the late nineteenth century produced—alongside the development of highbrow realist fiction—numerous popular fictions, often romances, that continued to imply the author's immanent personality.[23]

Hochman's account of the fortunes of the "friendly" narrator revolves around the American scene, but we can track similar shifts in British-authored texts. These also of course tended to chummy narration during the nineteenth century. According to Patrick Brantlinger, though, the growth of literacy in Victorian Britain, coupled with the "capitalization and industrialization of publishing," bred a lack of confidence in the reportedly indiscriminate, pleasure-seeking novel-consumer, making novelists throughout the period apprehensive about their audience.[24] Thus, too, Garrett Stewart asserts the contortion and eventual fading, especially in Modernist texts, of the Victorians' familial direct address due to authors' increasingly uneasy attitude toward the reading public.[25] By the late 1800s, industrial and democratizing forces were exerting a decisive pressure on literary forms, and high art was progressively cordoned off from the low.

Hochman points to *Trilby*, which had its first big success in the United States as a serial in *Harper's Monthly*, as an example of a work that maintained the trope of friendly narration and thereby offered a popular alternative to more off-putting literary styles. The degree of intrusiveness of the book's narrator is extraordinary, as she observes, generating a sense of community between Du Maurier and the reader. However, she sees a very different type of interaction with one's audience in the character of Trilby: because the heroine's mesmerism means she is never self-present while performing, her audience experiences nothing like the intimacy implied in the book's own figure of the "friendly author-narrator."[26] But I think the apparent contrast between *Trilby* the novel and Trilby the performer disappears once we view

23. Barbara Hochman, *Getting at the Author: Reimagining Books and Reading in the Age of American Realism* (Amherst: University of Massachusetts Press, 2001).

24. Patrick Brantlinger, *The Reading Lesson: The Threat of Mass Literacy in Nineteenth-Century British Fiction* (Bloomington: Indiana University Press, 1998), 13.

25. "Given the philistinism and *hypocrisie* of the mass *lecteur*, it came to seem beneath literature's dignity to enter into even the pretense of dialogue with such a constituency. The insistence on literature's inherently closed form can thus...be construed as a retreat from the mass contamination into art's impregnable fortress"; Garrett Stewart, *Dear Reader: The Conscripted Audience in Nineteenth-Century British Fiction* (Baltimore: Johns Hopkins University Press, 1996), 150–51. For Stewart's later analysis of *Trilby* as a staging of feminine emotional absorption in the act of reading, see 352–56.

26. Hochman, *Getting at the Author*, 68.

certain elements of the situation more closely. For one thing, to compare Trilby's performance to Du Maurier's authorship is misleading because Trilby is
not really analogous to an author—is not herself a creator. Rather, she serves
merely as a creator's (Svengali's) medium. For another thing, although Hochman's hyphen conflates them, an author is not necessarily identical with his
or her omniscient narrator, even when the author uses the first person to
give the appearance that this identity exists. Instead, Trilby and the friendly
narrator are both mediating figures that produce a certain artistic effect—
both instruments that transmit, like the phonograph or typewriter, but also
crucially construct the "message"—and that effect is a sense of emotional
togetherness that facilitates a popular reception.

Importantly, even if Trilby lacks self-presence while performing, her audience cannot perceive this; for them she really *is* looking back at them with
her hand to her heart, reaching out to them through her song, and it is in
large part this intimacy that entrances them.[27] Like other writers of his day,
Du Maurier underscores the mediating woman as a carrier of sympathy:
"But her voice was so immense in its softness, richness, freshness...and the
seduction, the novelty of it, the strangely sympathetic quality!" (191). Once
she has "saddened and veiled and darkened her voice," even an "absurd old
nursery rhyme" becomes for her listeners "the most terrible and pathetic
of all possible human tragedies, but expressed...by mere tone, slight, subtle
changes in the quality of the sound—too quick and elusive to be taken count
of, but to be felt with, oh, what poignant sympathy!" (191–92). Much of the
account of Trilby's Parisian concert dwells on the various states of breathless
compassion she inspires in her listeners en masse, often for the characters
about whom she sings. Her rendition of "Malbrouck s'en va-t'en guerre,"
for instance, contains a "note of anxiety...so poignant, so acutely natural and
human, that it became a personal anxiety of one's own, causing the heart to
beat, and one's breath was short" (196). The pathos of Trilby's song and of her
"dove-like eyes" (192) is so potent that it suddenly cures Little Billee of his
strange affliction of emotional indifference: "something melted in his brain,
and all his long-lost power of loving came back with a rush" (193). But the
pathos itself has a spurious source. Hardhearted as the novel shows him to be,
Svengali certainly does not share his audience's sentiments, even while engi-
neering his medium's affect to produce the greatest commercial impact.

Something similar is going on in Du Maurier's relation to his readers, although that situation may not have fully manifested itself until after the fact

---

27. Thus the quick-selling stereoscopic photographs of Trilby become a metaphor for her onstage demeanor: just as the stereoscope—a viewer through which an image appears to take on three-
dimensional form—creates the illusion of depth out of flatness, so does her mesmeric performance.

of the book's success. On the one hand, his friends remarked with pleasure on the resemblance of *Trilby*'s narrative voice to his own manner of talk, a resemblance that gave the book an "easy conversational style" that his biographer Leonée Ormond proposes proved highly attractive to a general readership.[28] Reviewers frequently commended his novel for what one called its "kindliness and sympathetic quality,"[29] terms that strongly mirror the book's description of Trilby's song. (In the same vein, her performance of the nursery rhyme, like her one of "Malbrouck s'en va-t'en guerre," renders in miniature *Trilby* as a whole: each begins lightheartedly but becomes gradually more grave, shifting from comedy to tragedy through tonal modulations.) On the other hand, as James made clear in an essay published shortly after Du Maurier's death in 1896, Du Maurier despised the popular clamor that beset him after his success as a novelist. "Why," James wonders, in language that vicariously suggests his friend's disgust, "did the public pounce on its prey with a spring so much more than elephantine?" The craze reached such proportions that it "darkened all [Du Maurier's] sky with a hugeness of vulgarity."[30] This is not simply one author imposing his own aloofness on another: Du Maurier's increasing publicity did depress him, exacerbating his illness, and a steady deluge of fan letters "made him angry and resentful."[31] So even as critics were observing of *Trilby*, "It is a story written out of the author's very heart, and it finds its way straight to the hearts of his readers" and calling this the "secret of its unique success,"[32] Du Maurier's private reaction to the attention of the masses betrays a profound emotional disconnection.

It is a lingering question how much Du Maurier himself understood the lucrative potential of and deliberately fabricated, as Svengali does, an openhearted mediating voice. Of relevance here are James's memories of attending a lecture by Du Maurier on *Punch* illustrators, part of a tour Du Maurier began in the early 1890s, just before his drafting of *Trilby*. What most impresses James about this lecture in retrospect is that Du Maurier seems to have "lighted on" just the right manner for "addressing the many-headed monster" that was his audience:

He had just simply found his tone, and his tone was what was to resound over the globe; yet we none of us faintly knew it... As this tone, I repeat, was essentially what the lecture gave, the best description of it

---

28. Ormond, *George Du Maurier*, 459.
29. Quoted in Purcell, "Trilby and Trilby-Mania," 68.
30. Henry James, "George Du Maurier," *Harper's New Monthly Magazine* 95 (1897): 604, 607.
31. Ormond, *George Du Maurier*, 462.
32. From the preface to *Trilbyana*.

is the familiar carried to a point to which, for *nous autres,* the printed
page had never yet carried it. The printed page was actually there, but
the question was to be supremely settled by another application of it.
It is the particular application of the force that, in any case, most makes
the mass (as *we* know the mass,) to vibrate; and Trilby still lurked unseen behind... Ibbetson... The note of prophecy, all the same, had been
sounded... The game had really begun, and in the lecture the ball took
the bound that I imperfectly indicate. Yet it was not till the first instal-
ment of *Trilby* appeared that we really sat up.[33]

James accentuates Du Maurier's "familiar" "tone" in the lecture as a "prophecy" of the style that would later ensure *Trilby*'s ability (figured in a way
reminiscent of the mesmeric rapport) to make "the mass... vibrate." But
noteworthy in this context is the revelation that this seeming affability could
coexist with basically mercenary motives: prior to the passage quoted above,
James has said that the lecture tour was primarily a money-making venture
that Du Maurier undertook for the support of his family and "loved... as
little as possible." James claims in the course of his anecdote that Du Maurier
was unaware of the particular aptness of his tone, but would this really have
been the case? After all, Du Maurier had by this time shown himself in his
cartoon work for *Punch* to be sharply perceptive of social desires, whims,
and conventions, and it is also worth recalling his penchant for lampooning
them.[34] At the limit we can conjecture that his mechanically smiling diva
mocks the reality that entertainment with mass appeal requires not so much a
genuine community between artist and audience as the performance through
some convincing medium of such a consensus.[35] *Trilby*'s image of the frenetic
reception of the singer's audience already implies this parodic take on the
popular.

So does the note of self-consciousness that begins the narrator's account
of Trilby's free-loving past:

She had all the virtues but one; but the virtue she lacked... was of such
a kind that I have found it impossible so to tell her history as to make

---

33. James, "George Du Maurier," 603.

34. See T. Martin Wood, *George Du Maurier, the Satirist of the Victorians: A Review of His Art and
Personality* (London: Chatto and Windus, 1913) and Ormond, *George Du Maurier,* chap. 8.

35. This secret of the novel brings to mind a tension Ivan Kreilkamp notices within Victorian
fiction, between a mythically authentic, humane storytelling voice and the modern developments
(including the invention of the phonograph) that threatened to erode this voice. See *Voice and the
Victorian Storyteller* (Cambridge: Cambridge University Press, 2005).

it quite fit and proper reading for the ubiquitous young person so dear to us all.

Most deeply to my regret. For I had fondly hoped it might one day be said of me that whatever my other literary shortcomings might be, I at least had never penned a line which a pure-minded young British mother might not read aloud to her little blue-eyed babe as it lies sucking its little bottle in its little bassinette.

Fate has willed it otherwise.

Would indeed that I could duly express poor Trilby's one shortcoming in some not too familiar medium—in Latin or Greek, let us say—lest the Young Person (in this ubiquitousness of hers, for which Heaven be praised) should happen to pry into these pages when her mother is looking another way.

Latin and Greek are languages the Young Person should not be taught to understand—seeing that they are highly improper languages, deservedly dead—in which pagan bards who should have known better have sung the filthy loves of their gods and goddesses. (32–33)

The readership is assumed here to be feminine (daughters or young British mothers), menacingly widespread (characterized by "ubiquitousness"), and ignorant of high culture (immune to Latin and Greek), and the narrator's exaggerated wish to satisfy them gives the passage all its irony. As Du Maurier implies, one must sometimes go to ridiculous lengths, such as writing in a dead language, to find a "medium" for telling the story that will cater to the straitened sensibilities of a bourgeois audience. The ironic tone likely would not have escaped his readers, who may have even found it amusing, but that effect would have only redoubled the satirical paradox. The narrator exhibits not only sensitivity but also charm, all the while deriding the necessity for doing so.

Du Maurier's use of narrative voice and style as an affective common ground with the reader recalls Marie Corelli's self-identification with *A Romance of Two World*'s first-person narrator, which she uses to speak more directly to her audience and thereby to cultivate a "sympathetic union." In these two authors, we see applications of the narrator's position as a medial one, as a means for an emotional extension connecting author to reader and a way to craft reader response. We also see that authors can implement this tool more or less opportunistically. It is noteworthy that two writers whose careers begin to carve out modern notions of the bestseller depict mediating acts within their works that remind us of the linguistic mediation of the works themselves, a mediation helpful to their popularity. And that the narrator's channeling role is doubled and personified within both writers' stories by female characters points up the perception that communicating feeling and intention between parties is best reserved for women.

False Womanhood and the Popular Medium

Whereas Corelli focuses on the sensitive channel as an authentic moral agent, Du Maurier focuses on her manufacture: his portrait of La Svengali highlights the gap between what she seems to feel for her audience and the secret dishonesty of that feeling. This fraud—another instance of the automatizing villain's defilement of his victim's capacity for sympathy—translates as a perverted femininity. From her audience's perspective, the singer embodies femininity in what is reputedly its most caring state: her voice is like "a broad heavenly smile of universal motherhood turned into sound" (191).[36] But what they do not know is that at bottom this motherhood is a horrific parody: the only thing phonographic La Svengali really reproduces is Svengali. Du Maurier drives the point home in one of his final illustrations for the novel (fig. 1). Trilby has just received Svengali's photograph in the mail, and though he has already died, his image propels her into a trance, complete with a rendition of Chopin's Impromptu in A flat (259–61). The theme of reproducing Svengali, already implied in the plot detail of his photograph, is reinforced by Du Maurier's caption, which renders Trilby's reiterative trance utterance: "Svengali!...Svengali!...Svengali!" That Trilby, not the camera, has been the principal recording and reproducing instrument of Svengali is stressed by her facial expression and physical position: while these presumably depict her mesmeric rapture, they also call to mind a woman in the agonies of childbirth. Her clustering friends are powerless to stop the debilitating song generated by Svengali's psychic rape.

La Svengali's make-up also indicates the mendacity of her feeling. Even as her voice explodes with maternity, her reddened lips and cheeks hint at a merely "artificial freshness" (189). Later, when Little Billee spies her on the street, she is "rouged and pearl-powdered, and her eyes...blackened beneath, and thus made to look twice their size" (213). She wears sables, yet cuts him with a "little high-pitched flippant snigger worthy of a London barmaid" (213). Like the comparison to the barmaid (which James's In the Cage also

---

36. Elaine Showalter reads this description as indicating Svengali's project of taking hold of a woman's "creative organs of maternity"; introduction to Trilby, by George Du Maurier (Oxford: Oxford University Press, 1998), xx.

FIGURE 1. "Svengali!...Svengali!...Svengali!" Trilby has just viewed Svengali's photograph and is entranced one last time. George Du Maurier, *Trilby*, 1894.

exploits), the clash between La Svengali's finery and her coarse laugh conveys the meretriciousness of the prostitute, and in turn her empty affections. As opposed to the original Trilby—who loves if too bodily and freely, at least "for love's sake only" and with an honest "gaiety of heart and genial good-fellowship" (33)—automatized Trilby only imitates love, for a price.

Mesmerized Trilby's deceptive shows of love make her, symbolically, the most stigmatized of Victorian women—a sham mother; a whore—with the tendentiousness of the symbolism registering how much loving sympathy was revered as a feminine trait. But that kind of gendering does not stop with the Victorians, and by extension we can hear the strains of *Trilby*'s suspicious representation of mass media echoed in later writing. There is something suggestive of a reified intimacy or sympathy in Walter Benjamin's concept of aura, an idea of a community of the viewer with the artwork and, through it, with humanity and tradition across the ages. (And this is true whether that affirmation of tradition is valued positively or negatively.)[37] This is presumably what Benjamin means when he states that auracular artwork seems to

---

37. For a discussion of aura as an agent of human empathy and community, and of Benjamin's alternate longing for it and welcome of its loss in the face of less fascistically traditional forms of art, see Andrew Arato and Eike Gebhardt, eds., *The Essential Frankfurt School Reader* (New York: Continuum, 1982), 209, and generally 208–12.

gaze back: the gaze is really the viewer's seeming access by means of the work to a responsive human impulse from the past. By contrast, photographic art erodes the experience of seeming mutual recognition due to the interposition of a lifeless instrument: "What was inevitably felt to be inhuman, one might even say deadly, in daguerrotypy was the (prolonged) looking into the camera, since the camera records our likeness without returning our gaze."[38] For Benjamin as for Du Maurier, the failure of sympathy inherent in the process of mechanically reproduced art becomes a sexually loose but emotionally inaccessible woman. Benjamin insinuates that idea analogically through his explication of the poetry of Baudelaire, who reportedly dwells on the difference between, on the one hand, an intimate gaze and, on the other, an only empty stare:

> In eyes that look at us with a mirrorlike blankness the remoteness remains complete. . . . Baudelaire incorporated the smoothness of their stare in a subtle couplet:
>
>> Plonge tes yeux dans des yeux fixes
>> Des Satyresses ou des Nixes.
>> [Let your eyes look deeply into the fixed stare
>> Of Satyresses or of Nymphs.]
>
> Female satyrs and nymphs are no longer members of the family of man. Theirs is a world apart. . . . [The poet] has lost himself to the spell of eyes which do not return his glance and submits to their sway without illusions.[39]

The "sway" of the satyresses and nymphs is of course their sexual allure—a poor substitute for authentic sympathy (a place within the "family of man"). The utter "remoteness" of their eyes whets desire while precluding the possibility of genuine affective fulfillment.[40] The only time the deadened eyes of these women "come alive, it is with the self-protective wariness of a wild animal hunting for prey. (Thus the eye of the prostitute scrutinizing the passers-by is at the same time on its guard against the police.)"[41]

---

38. Benjamin, "On Some Motifs," 187–88.

39. Ibid., 190; translation in original.

40. It is worth clarifying that Benjamin seems to use the term "distance" differently. The auracular artwork exhibits a distance from the viewer by virtue of its "ceremonial" aspect—its association with the sacred and the traditional—yet Benjamin implies that that distance is in some sense bridgeable by sympathetic interconnection. Total "remoteness" is, by contrast, unbridgeable. See ibid., 188–90.

41. Ibid., 190–91.

In such a context, it is not enough to say that mass culture is a woman. Mass culture is a *false* woman: one whose demonstrations of love are superficial and sordid. By definition, popular art unifies, producing a state resembling sympathy among its audience. Yet the feelings generated in this experience need not refer back to a sincere human origin but are rather a potentially rootless, illusory effect produced by modern media themselves—their powers of imitation, manipulation, and dissemination. Furthermore, the unification of the audience through the work is likely to be ephemeral, as much so as a harlot's tryst, giving way easily to other sources of gratification, for instance the next bestseller. One could point to any number of ingredients in the production of mass art as they contribute to this situation, but Benjamin's and Du Maurier's writings both play up the specious charm of the medium it-/herself.[42]

Yet finally it is worth emphasizing that *Trilby* is also particular to its historical moment. Even if the novel's depiction of reproductive media resembles Benjamin's, the two perspectives also fundamentally differ for the basic reason that Du Maurier is writing some forty years earlier, when these technologies were less common and, moreover, easily yoked to interests in the occult, which obviously color the major drama of his novel. *Trilby* is caught between, on the one hand, a prescient sensitivity to how a medium like the camera underwrites worldly commercial ventures (like those of the Stereoscopic Company) and, on the other hand, an awe at its seemingly phantasmal creations. Recall in this light the episode in which Trilby receives Svengali's photograph. As the narrator tells us, the mesmerist seems to be "looking straight out of the picture, straight at you," with his "big black eyes... full of stern command" (259). Is this another case, as with Trilby onstage, of a medium whose fake gaze only seems true? Or has this photograph somehow really attained to the status of auracular art?[43] Probably neither is quite accurate. I think it is most likely that the camera has simply managed to transcend the boundary between life and death. In shifting in this scene from Trilby's phonographic reproduction of Svengali to a photographic one, Du Maurier takes advantage of the camera's Victorian reputation for capturing the

---

42. I mean to describe the seeming harlotry of mass media or communication in particular, rather than of modern technology in general. For the latter argument, see Huyssen, who claims that Fritz Lang's *Metropolis* figures technology as a vamp in order to convey men's anxieties about their ability to control it, *After the Great Divide,* chap. 4. See my reading of Rudyard Kipling's "Wireless" in chap. 5 for a related exploration of eroticized electricity and male responses to it.

43. Interestingly, Benjamin's theory allows for this possibility, in his mention of early portrait applications of photography: "The cult of remembrance of loved ones, absent or dead, offers a last refuge for the cult value of the picture. For the last time the aura emanates from the early photographs in the fleeting expression of a human face" ("Work of Art," 226). But the aura of photography dwindled, he says, as portrait photography lost ground to other uses of the medium.

departed. Just as "spirit photographs" when developed suddenly seemed to exhibit ghostly contours, this picture renders Svengali's gaze so compellingly because he himself really *is* occultly present in this artifact, entrancing Trilby from beyond the grave.[44]

Importantly, he is able to do here what—postmortem then as well—he failed to do during Trilby's last debacle of a concert. There, although seated in a box directly in front of her and staring back at her, he could not induce her to sing one good note. But here in the photograph, he achieves all the glory of her Chopin's Impromptu. The difference in effects conveys, in the first place, Svengali's preference, as a cultural outsider who repels in person, for mediated acts of captivation: he has always worked most spectacularly by working *through*—in this case through the photograph. But there is, further, the implication of some special power in the camera itself to lay hold of personality in a way the mortal body cannot. At the turn of the century, reproductive media facilitated theories that defied a mundane understanding of humans' existential and perceptual limits. And if *Trilby* portrays terrifying consequences of such scenarios, Du Maurier's other novels offer brighter stories that reinforce hopes for spiritual permanence.

The Ghost in/and the Machine: Technologies of the Unconscious Self

Trilby's literary antecedent, the android Hadaly of *L'Eve future,* begins to illustrate these occultist valences of fin de siècle reproductive technologies. Besides Hadaly, Edison houses at Menlo Park a woman named Anny Anderson, the widow of a longtime friend. Hypnotized by Edison to cure her of a neurosis, Mrs. Anderson has developed paranormal powers, including clairvoyance, that are inexplicable to him and that may have even permitted her to transfer an ethereal part of herself—a second, hypnotic-state personality calling itself Sowana—into Hadaly's mechanical body. In a private conversation with Ewald near the end of the novel, the now reconstructed android intimates this transfer and, with it, a metaphysical basis to her being. Alicia-Hadaly describes herself as an "envoy... from limitless regions," declaring about Edison, "I called myself into existence in the thought of him who created me, so that while he thought he was acting of his own accord, he

---

44. On spirit photography, see the photographic history in Clément Chéroux et al., *The Perfect Medium: Photography and the Occult* (New Haven: Yale University Press, 2005).

was also deeply, darkly obedient to me" (198). Thus ultimately the novel puts forth two possible accounts for the verbalizations of Edison's android: Hadaly may only be playing back, as intended, the scripted words of geniuses, but alternatively, at least some of her words may have their source in an otherworldly realm. The android's phonographically centered self is complicated, because spiritualized, by the latter possibility. Her recording cylinders, though painstakingly patterned by Edison, might finally be subordinate to an identity that exceeds his materialist efforts.

In his analysis of *L'Eve future,* Friedrich Kittler overlooks this implication that the android's speeches to Lord Ewald originate in some authentic spirit. His reading insists instead on Edison's victory in mechanically simulating the Ideal, which he says male poets and philosophers living a century before Edison had envisioned as a transcendent signified of Nature embodied in a Woman who inspired their writings: "[Villiers's] technological substitute perfects *and* liquidates all the characteristics attributed to the imaginary image of Woman by Poets and Thinkers."[45] Indeed, it is not surprising that Kittler accentuates Edison's apparently consummate work of artifice and, conversely, suppresses the possibility of Hadaly's spiritually originating utterances. To admit the latter would be to admit what his assessment of the "discourse network" circa 1900 largely denies: that technologies of writing and verbalization were still regarded by intellectuals of that era, as they had been by the Romantics circa 1800, as capable of communicating a transcendent and meaningful Ideal.

But it is just this Ideal that Villiers labored to communicate in his own writing, an attempt well understood by a likeminded contemporary, Symbolist Arthur Symons:

> The ideal, to Villiers, being the real, spiritual beauty being the essential beauty, and material beauty its reflection or its revelation, it is with a sort of fury that he attacks the materialising forces of the world... Satire, with him, is the revenge of beauty upon ugliness, the persecution of the ugly. It is not merely social satire, it is a satire on the material universe by one who believes in a spiritual universe.... And this lacerating laughter of the idealist is never surer in its aim than when it turns the arms of science against itself, as in the vast buffoonery of *L'Eve Future.*[46]

---

45. Friedrich Kittler, *Discourse Networks, 1800/1900,* trans. Michael Metteer, with Chris Cullens (Stanford: Stanford University Press, 1990), 348.

46. Arthur Symons, "Villiers de l'Isle Adam," *Fortnightly Review* 66 (1899): 201.

By its mystical final scenes, Villiers's novel ends not by celebrating but rather by pointing up the deficiencies in Edison's technological response to Ewald's quest for love, human fulfillment, and the Ideal. When Edison hypnotizes Mrs. Anderson, he unwittingly opens the door to a world of spirit and wonder at best sketchily apprehended by his science. The novel's "buffoonery" lies in its last-minute collapse of Mrs. Anderson's soulful unconscious with the automaton's recording cylinders; the result is a weirdly hybrid being who refutes the materialist philosophy of Hadaly's "creator."

Kittler is probably drawn to a study of Villiers's android because she appears prima facie to substantiate his claim that turn-of-the-century technologies ushered in a cultural shift toward visions of spiritless, mechanically derived human utterances. Devices like Edison's phonograph, Kittler claims, combined with two contemporary sciences—psychoanalysis and psychophysics (the examination of the effect of physical stimuli on cognitive and perceptual states)—to reduce images of mental processes, and the words and writing produced from them, to merely material operations uninformed by an underlying ontological identity. In one example, Kittler looks at a philosopher's analogy between the phonograph and memory and draws from it a notion of the brain as essentially an etched cellular document of lived experience: "All questions concerning thought as thought have been abandoned, for it is now a matter of implementation and hardware."[47] Alternatively the phonograph might serve externally in scientific research, collecting utterances that are more "noises" and "linguistic hodgepodge" than intentionality; thus Kittler likens phonographic recordings to the real of Lacan.[48] According to such readings, the late nineteenth century boiled down ostensible expressions of personality to neural functions, psychoanalytically legible libidinal drives, and so on, thoroughly repudiating a Romantic picture of innerness. The phonograph—which "means the death of the author" as "it stores a mortal voice rather than eternal thoughts and turns of phrase"—coincides with theories of the mind's productions as devoid of any fundamental, let alone spiritual, intelligence or selfhood.[49]

But what becomes clear from examining turn-of-the-century literature and culture is that Kittler's reading excludes much, and much of importance, in its narrow focus on psychophysics and psychoanalysis as the methods by

---

47. Friedrich Kittler, *Gramophone, Film, Typewriter,* trans. Geoffrey Winthrop-Young and Michael Wutz (Stanford: Stanford University Press, 1999), 33.

48. Kittler, *Gramophone,* 85, 86. For his theory of Lacanian correspondences, see *Discourse Networks,* 246.

49. Kittler, *Discourse Networks,* 237.

which the mind was theorized in the discourse network of 1900. Another method, hypnosis, which Freud renounced early on in favor of the talking cure,[50] achieved a far greater hold on the imagination of the transatlantic public. From periodical essays and medical case studies to novels like *Trilby*, tales of hypnosis piqued a widespread interest and as a result contributed considerably more significantly than either psychophysics or psychoanalysis to the period's speculations about the productions and memories of the (unconscious) mind. These tales commonly emphasized not the dead materiality of neural physiology but instead a coherent intelligence at work on or even, phantomlike, *within* the subject's unconscious. Further, in some cases, the hypnotic paradigm entailed testimonies of extrasensory communication—and not simply in works of fiction like *L'Eve future*. Psychical researcher Frederic Myers was hardly alone in his day in yoking such phenomena to some unconscious, spiritually attuned aspect of the self. It is especially this aspect that Kittler's reading of turn-of-the-century discourse effectively, while not directly, disclaims, but which I think an earnest look at contemporary hypnotic and related theories of mind will allow us to recover.

In addition, it is evident that whatever their role in psychophysical experiments, recording and reproducing technologies, the phonograph as well as the camera, assisted writers in positing the occult potentialities of the unconscious. Algernon Blackwood's "The Transfer" (1912) associates the camera with clairvoyance, paralleling a governess's preternatural vision with the development of a photograph. Another Blackwood story, "If the Cap Fits—" (1914), characterizes a psychometrist (a person gifted with the power to perceive historical moments simply by touching objects associated with them) as a "sensitive photographic plate."[51] These kinds of comparisons seem indebted to a nineteenth-century axiom—useful to believers in spirit photography as well—that the camera had a privileged access to truth and could capture even realities that escaped average human perception.[52] For late Victorian spiritualist and journalist W. T. Stead, the power of even a primitive

---

50. Henri Ellenberger, *The Discovery of the Unconscious: The History and Evolution of Dynamic Psychiatry* (New York: Basic Books, 1970), 519, 802.

51. Algernon Blackwood, "The Transfer," in *Best Ghost Stories of Algernon Blackwood*, comp. and intro. E.F. Bleiler, 228–39 (1912; New York: Dover, 1973); Algernon Blackwood, "If The Cap Fits—," in *Ten Minute Stories* (1914; Freeport: Books for Libraries, 1969), 221.

52. On the Victorian faith in the camera's revelatory power, see Jennifer Green-Lewis, *Framing the Victorians: Photography and the Culture of Realism* (Ithaca: Cornell University Press, 1996). The "Coda" discusses spirit photography specifically. This is not to say that the relationship between photography and "reality" was uncontested or always conceived in the same way. Other concepts of photographic realism turned on the manipulation and reconstruction of photographs; see Daniel A. Novak, *Realism, Photography, and Nineteenth-Century Fiction* (New York: Cambridge University Press, 2008).

form of photography, or light-writing, to offer phantasmal reproductions helped to suggest unconscious links to a numinous domain. Having already described the unconscious as a buried area of intelligence expertly penetrated by the hypnotist, Stead declares its mystical powers by theorizing a cosmic entity he dubs the "astral camera" or an "invisible camera obscura in Nature." The age-old medium of the camera obscura is a dark box or chamber penetrated by a hole through which light streams, replicating the external world. It brings remote scenes near, as do clairvoyants in Stead's analogy. An all-seeing perceiver of lived events, his astral camera is accessible to the unconscious of clairvoyants and other paranormally receptive people: "In the mysterious, subconscious world in which the clairvoyant lives, may there not be some subtle, sympathetic lens . . . which may enable some of us to see that which is quite invisible to the ordinary eye?"[53] Psychometrists also rely on the astral camera; yet here, interestingly, Stead subtly conflates his older model of the camera obscura with the photography of his own century, when Louis Daguerre for the first time chemically fixed images, granting the technology the ability to record: "This astral camera, to which 'future things unfolded lie,' also retains the imperishable image of all past events . . . Any article or relic may serve as a key to unlock the chamber of this hidden camera."[54]

Stead imagined that the unconscious of the psychometrist tapped into a mnemonically capacious technology; contemporary researchers of the mind variously implied that the unconscious itself was such a technology.[55] In *The Psychology of Suggestion: A Research into the Subconscious Nature of Man and Society* (1898), American psychologist and hypnotist Boris Sidis reports the work of a colleague who found that in pressing his patient's neck, "echolalia resulted—the patient repeated everything that was said before him with the exactness of a phonograph."[56] Sidis was no believer in spirits; even so, the recording unconscious in his work intimates a normally hidden yet discrete stratum of selfhood that is vaguely spectral. His own hypnotic encounters with his patients also yield a kind of cerebral playback, but in describing them he deviates from his colleague's findings of merely disjointed, unthinking vocalizations. For Sidis places the origin of the playback not in the doctor's arbitrary words but in the patient's own comprehensive memory: the patient's mind reproduces data lost to consciousness yet recorded within the unconscious mind.

---

53. W. T. Stead, *Real Ghost Stories* (1891; New York: Doran, 1921), 113, 110.
54. Ibid., 126.
55. For another consideration of the phonograph and photography as mnemonic models, see Douwe Draaisma, *Metaphors of Memory: A History of Ideas about the Mind*, trans. Paul Vincent (Cambridge: Cambridge University Press, 2000), 90–93 and 119–34, respectively.
56. Sidis, *Psychology of Suggestion*, 80.

One of the book's anecdotes to this effect concerns Sidis's widely discussed patient Reverend Thomas C. Hanna, a man who suffered a fall that left him amnesiac as to his own identity and even language and other basic capacities. However, Hanna's amnesia resided solely in his conscious mind; occasionally his unconscious managed to bring forth—verbatim, phonographically—the knowledge it had preserved:

> Events, names of persons, of places, sentences, phrases, whole para-graphs of books totally lapsed from memory, and in languages the very words of which sounded bizarre to his ears and the meaning of which was to him inscrutable—all that flashed lightninglike on the patient's mind. . . . To the patient himself it appeared as if another being took possession of his tongue.[57]

Here, as when commenting on *Trilby*'s popularity, Sidis uses an idiom of "pos-session" to explain the marvels uncovered by studies of hypnotic suggestion; in this case the image is clearly of an occult being lodged within or identical with the patient's unconscious mind. Notably, *The Psychology of Suggestion* earlier characterizes the unconscious as "an intelligent, subwaking, hypnotic self concealed behind the curtain of personal consciousness," a phrasing that recalls the spirits and spiritual effects often alleged to materialize from behind the curtained-off spaces of séance mediums.[58] All in all, Sidis's idea of the un-conscious—that in hypnotic speech and writing it exhibits a true "self, pos-sessing consciousness, memory, and even a rudimentary intelligence"[59]—is a far cry from the producer of stochastic refuse in Kittler's schema; it is much closer to how Stead envisioned it: as a "ghost that dwells in each of us."[60]

The tacit parallel with an apparition suits Sidis's meaning because he theo-rized the unconscious as a second, usually obscured region of real conscious-ness—and not, as Freud did, a fact of literal non-consciousness. This former notion Adam Crabtree calls the alternate-consciousness paradigm, and it is one that Sidis shared with other prominent psychologists of his day, includ-ing Pierre Janet and Morton Prince.[61] The hidden train of consciousness

---

57. Ibid., 224.

58. Ibid., 100. Note for instance how often curtained séance spaces figure in the photographs of mediums in Chéroux et al., *Perfect Medium*.

59. Sidis, *Psychology of Suggestion*, 106.

60. This is the title of Part I of Stead's *Real Ghost Stories*.

61. As Crabtree remarks, Janet and Prince shied away from the term *unconscious* and veered toward *subconscious* and *co-conscious*, respectively, to make clear their hypothesis of an authentic but buried consciousness. Adam Crabtree, *From Mesmer to Freud: Magnetic Sleep and Roots of Psychological Healing* (New Haven: Yale University Press, 1993), 317, 349.

was seen as accessible through hypnosis; in fact, Crabtree shows, hypnosis and its antecedents are the historical core of the paradigm because they seemed to unearth trains of thought and memory manifestly separate from the waking self.[62] In cases of hysteria, a second consciousness might develop so independently of the main consciousness as to rival with it for bodily control, transforming into a walking, talking second personality. It is the overall conception of the unconscious as alternate consciousness that, here and elsewhere in this book, I am concerned with. Though the unconscious, like hysteria, has tended to be treated as Freud's special province and discovery, psychoanalysis hardly emerged in a vacuum but rather displaced certain dominant ways of viewing buried mental life, concentrated at the turn of the century in writings around hypnosis and paranormal practices. Studying trance and the various types of covert selves it revealed gives us an on-the-ground view of how the mind seemed to operate and communicate and of how media and mediation seemed to play a part in these communications.

In their efforts to understand aberrant mental occurrences, both psychologists and psychical researchers were fascinated by the secret workings of the unconscious. Psychologists interested in the topic often drew on the same archive as did psychical researchers: observations around hypnotic phenomena and (purported) spirit mediumship. This made for considerable theoretical and practical overlap, and indeed, psychologists sometimes acknowledged psychical researchers as important fellow travelers on the same road to knowledge.[63] Edmund Gurney, one of the co-founders of the SPR, divided his time between his researches into phenomena like telepathy and his hypnotic experiments; his psychical research is sometimes indistinguishable from his work in psychology, as is also true of Frederic Myers.[64] Particularly notable is Myers's concept of the "subliminal" self, which he saw as a vast area of consciousness—in fact multiple streams of volition, thought, intention, and memory—beneath the normal or "supraliminal" self. Aspects of the subliminal are occasionally capable

---

62. Freud eventually diverged from the theory of secondary hypnoid states of Josef Breuer, his co-author on *Studies in Hysteria,* as he began to turn toward a theory of consciousness as unitary, in which much mental life falls outside of its "searchlight" beams; Crabtree, *From Mesmer to Freud,* chap. 17, esp. 357–60. Alan Gauld elaborates on Myers's and others' difference from the searchlight model in *The Founders of Psychical Research* (New York: Schocken Books, 1968), 278–80.

63. On psychical researchers' and psychologists' interests in matters of the unconscious, see Crabtree, *From Mesmer to Freud,* chaps. 14–17. Much important turn-of-the-century psychological study of the complex self emerged from France; John Warne Monroe, *Laboratories of Faith: Mesmerism, Spiritism, and Occultism in Modern France* (Ithaca: Cornell University Press, 2008), chap. 5, discusses how French Spiritism, Occultism, and psychical research dovetailed with, and diverged from, psychological research.

64. Oppenheim, *The Other World: Spiritualism and Psychical Research in England, 1850–1914* (Cambridge: Cambridge University Press, 1985), 120.

of breaking through the barrier of the supraliminal; thus Myers accounted for a wide range of psychical effects—from dreams, hallucinations, hysteria, and flashes of genius to telepathy, clairvoyance, and spiritualistic contact. Certain strange aptitudes of the subliminal were for Myers evidence of our cosmic evolution and immortality. Such transcendental elements of the theory inspired some misgivings in workers in the burgeoning psychological field. Nonetheless, his basic framework of the subliminal was taken seriously, and he earned the admiration of a number of international psychologists, whose work he observed firsthand and with whom he sometimes collaborated.[65]

The interchange of ideas between psychical researchers and psychologists helps to explain the occultist rhetoric that occasionally inflects the latter group's writings. At moments, it is somewhat difficult to differentiate psychologists' methods for investigating the abnormal from others' approaches to the paranormal. Sidis describes probing his patient's unconscious with some noteworthy aids: "Now all these facts of crystal-gazing and shell-hearing clearly reveal the presence of a secondary, submerged, hyperaesthetic consciousness that sees, hears, and perceives what lies outside the range of perception of the primary personal self."[66] Crystal-gazing and shell-hearing were regular practices among clairvoyants: they would concentrate on a globe or shell, from which otherwise imperceptible sights or sounds would then emerge. But as Sidis's work shows, these practices might also be applied to recover material forgotten by consciousness yet retained in the unconscious or "secondary" self. That self is all-recording, "hyperaesthetic," and its contents can be externally reproduced visually, like a camera's, or aurally, like a phonograph's. Discussing clairvoyance, Stead, for his part, cannot miss the analogy between the crystal vision and the photographic image:

> If you have the faculty, the glass will cloud over with a milky mist, and in the centre the image is gradually precipitated in just the same way as a photograph forms on the sensitive plate. At least, the description given by crystal-gazers as to the way in which the picture appears reminded me of nothing so much as what I saw when I stood inside the largest camera in the world, in which the Ordnance Survey photographs its maps at Southampton.[67]

---

65. On Myers's theories and their reception, see Crabtree, *From Mesmer to Freud*, chap. 16; Oppenheim, *Other World*, 254–63; and Luckhurst, *Invention of Telepathy*, 107–12. Myers's fullest treatment of the subliminal appears in his *Human Personality and Its Survival of Bodily Death*, 2 vols. (New York: Longmans, Green, 1903).

66. Sidis, *Psychology of Suggestion*, 157.

67. Stead, *Real Ghost Stories*, 144.

True Dreams, Subliminal Records:
Intimations of Immortality in Du Maurier

George Du Maurier's first novel, *Peter Ibbetson* (1891), reflects the conflu-
ence at this time of various theories of the mind implying latent powers of
extraordinary memory and perception. The story begins in France, where
the title character, known in these early years as Gogo, enjoys a blissful childhood, often in the company of the little neighbor girl, Mimsey. But after his
parents' deaths, Peter must return to his homeland of England to live with his
odious Uncle Ibbetson. As a young man years later, Peter crosses paths with the
beautiful and kind Mary, Duchess of Towers, who shortly thereafter appears in
his dreams. A second waking encounter with Mary reveals her identity as the
grown-up Mimsey, and she and Peter are astonished to deduce that his dream
was a telepathic one experienced by her also. After a distressing rumor about his
parentage leads him to murder his uncle in a fit of rage, Peter is sent to prison,
yet is consoled there by nightly telepathic meetings with Mary. With her aid he
masters the skill of diving into the scenes of his unconscious memory—in effect, of hypnotizing himself—a skill she calls "dreaming true." In the commu-
nal territory of their dreams, the lovers are able to recall together—sensually
and three-dimensionally, in a sort of Victorian virtual reality—episodes from
their individual and shared histories, the recollections of which have been
meticulously stored in their unconscious minds.

If the link between the unconscious and recording technologies is only
implied in *Trilby*, it is explicit in this novel. Peter muses, "Evidently our brain
contains something akin both to a photographic plate and a phonographic
cylinder . . .; not a sight or a sound or a smell is lost; not a taste or a feeling
or an emotion. Unconscious memory records them all."[68] Photographic and
phonographic technologies were of course quite distinct: the first chemically imprinted a light pattern onto a plate, while the second used a point or
stylus to etch sound vibrations onto engravable material. That Du Maurier
would so easily collapse the two shows he is less concerned with technical
processes than with what these technologies loosely shared: an ability to record a moment for later access—that is, a relationship to time. Both forged
links between past and present and between present and future. In this way,
both ground what I see as *Peter Ibbetson's* main imaginative exploration: the
individual's selfhood as a function of time, of his or her placement on a personal or human continuum.

---

68. George Du Maurier, *Peter Ibbetson* (New York: Harper and Brothers, 1891), 227. All further
page references to *Peter Ibbetson* are parenthetical within the text.

One of the most vivid expressions of this theme is Peter's dual existence as both adult and child.[69] The sharpness of the duality becomes particularly evident when Peter reacts to his childhood self as observed within the "camera-obscura" of his dream world (215). He finds amazing how separated he, the adult dreamer, feels from Gogo: "most inexplicable, I saw it all as an independent spectator, an outsider, not as an actor going again through scenes in which he has played a part before!" (228). In some respects this temporal split mirrors the real-life medical case of the Frenchman Louis Vivé. The neurotic Vivé alternated between several personalities, each a regression to some point in his childhood. At the turn of the century—when the condition of multiple or dissociated consciousness (seen also in Villiers's Sowana) was firing the transatlantic imagination[70]—news of the Vivé case traveled quickly from France to England, where it was summarized, for instance, in Stead's *Real Ghost Stories,* within a chapter provocatively titled "Louis V. and His Two Souls."[71] One is tempted to read Peter Ibbetson's alternate, French, childhood self as a fictionalization of Louis Vivé's own fractured selves, especially since these were often conjured, as is Gogo, in a state of hypnotic sleep.

Some of the leading hypnotists of the period diagnosed hysterical symptoms, such as dissociated states, as tokens of forgotten traumas.[72] While not signs of trauma, Peter Ibbetson's dissociative dreams adhere to the temporal logic of hysteria, because the splitting of the self occurs as an upsurge of a personal past executed at the level of the unconscious. Like Louis Vivé's, Peter's unconscious souvenir represents a copy of his past experience, as exact as would be issued by the recording instrument with which he equates his unconscious mind.[73] Another fin de siecle fiction trading on the then-familiar

---

69. Richard Kelly reads duality as an organizing principle of the plot; for him Peter is a "split personality" because he is half-English and half-French, and because he divides his life between dream and reality. *George Du Maurier* (Boston: Twayne, 1983), 60–66.

70. Ellenberger calls the diagnosis of dissociated consciousness one of the "most discussed [problems] by psychiatrists and philosophers" by the last decades of the century (*Discovery of the Unconscious,* 126).

71. Stead, *Real Ghost Stories,* chap. 2. For modern-day accounts of the Vivé case, see Crabtree, *From Mesmer to Freud,* 302–6 and Ian Hacking, *Rewriting the Soul: Multiple Personality and the Sciences of Memory* (Princeton: Princeton University Press, 1995), chap. 12.

72. On Janet's views on past traumas as roots of hysteria, see Crabtree, *From Mesmer to Freud,* 322–26 and Ellenberger, *Discovery of the Unconscious,* 372–73. Past traumas also figured heavily in Jean-Martin Charcot's theories, although he envisioned these to be of a physical rather than psychical nature; see Michael Roth, "Hysterical Remembering," *Modernism/Modernity* 3, no. 2 (1996): 1–30.

73. A couple of other critics have analyzed the hysteric's relation to the past in terms of hypnotist Jean-Martin Charcot's photography of his patients. Roth likens spirit photography to Charcot's photography as both attempts to produce a material record of the abnormal and of the past (the deceased,

topic of dissociation also invites a comparison of a hidden self to a duplicating technology, in this case a camera. It is little coincidence that the double memorializing Dorian Gray's sins materializes as a picture.

Although in one sense Peter hardly differentiates between the camera and phonograph when he thinks about unconscious recording, in another sense his story, like *Trilby,* leans on the phonographic model in representing the wonders of the retentive subliminal mind as marvelous auditory reproductions. Peter thinks of musical strains as, like smells, "rare sublimaters of the essence of memory" (64) and asserts the "strange capacity of a melodic bar for preserving the essence of by-gone things, and days that are no more" (29). Then there is the vast scope of his memory of music itself: inside his head is always running the "ghostly music" of some song, a perfect reproduction of the original (104). As he later discovers, these inner harmonies—which he facetiously calls "unconscious musical cerebration"—have been psychically inherited from his great-grandfather (103). Additionally, Peter's dreams reveal an unconscious so aurally perceptive that it has kept track even of dialogue unheeded by him in his youth. Thus it is in unconsciously traveling back in time and eavesdropping on his childhood that he discovers a certain damning piece of information about his uncle:

> I had found this out by listening (in my dreams) to long conversations between my father and mother in the old drawing-room at Passy, while Gogo was absorbed in his book; and every word that had passed through Gogo's inattentive ears into his otherwise preoccupied little brain had been recorded there as in a phonograph, and was now repeated over and over again for Peter Ibbetson, as he sat unnoticed among them. (253)

Besides the homey French scenery of their childhood, Peter and Mary's dreams allow them to return to much more remote points, beyond the limit of their personal existences. Perhaps the most striking aspect of Du Maurier's science fiction-romance is the way it literalizes in full sensory detail a proposition common to turn-of-the-century views of the mind: that the unconscious is a site of regression to a less developed condition of humanity. (This is for

or the physical traumas Charcot believed to induce hysteria). Ulrich Baer argues that in Charcot's photographic studio, the shock of the flash simultaneously produced and recorded the hysterical symptom of catalepsy. Photography was thus caught up in not only the treatment of hysteria but the "poetics" of the disease itself, including its somatic "flashbacks" to past traumas; "Photography and Hysteria: Towards a Poetics of the Flash," *Yale Journal of Criticism* 7 (1994): 41–77.

example what Le Bon means when he asserts that the mass hypnotism of the crowd calls forth the "atavistic residuum of the instincts of the primitive man.")[74] Peter and Mary discover they have inherited "antenatal memories" stretching back not just to their grandparents and great-grandparents but as far as prehistoric times, such that they are just able to make out the dim shape of the mammoth, "blurred and indistinct like a composite photograph" (356, 368; fig. 2), and its human hunter. Peter recalls, "Mary firmly believed we should have got in time to our hairy ancestor with pointed ears and a tail, and have been able to ascertain whether he was arboreal in his habits or not" (369). The subliminal mind is a composite or amalgam of past experience; as if in some composite photograph (a double exposure), especially the kind used in so-called spirit photography, our forebears are ghostly presences, residing within us—psychically but also, *Peter Ibbetson* argues by extension, biologically.[75] Taking his dreams' graphic proof of inheritance to heart, Peter prompts the reader to "go forth and multiply exceedingly" but in doing so to "select the very best of your kind in the opposite sex for this most precious, excellent, and blessèd purpose," that "all your future reincarnations (and hers)" will be tinged with "well earned self-approval" (360). By the same token, the reader should avoid bad couplings: "beware and be warned in time, ye tenth transmitters of a foolish face, ye reckless begetters of diseased or puny bodies, with hearts and brains to match!" (361–62).

Peter's eugenicist philosophy substantially transmutes the novel's picture of unconscious devolution. Whereas the view of the mind as the repository of our primeval past might have encouraged a grim degenerationist outlook, Du Maurier's novel deploys it instead as a pointer toward the future and for a speculation as to how one's identity will accumulate over time. The reference to "diseased or puny bodies" notwithstanding, the emphasis is finally not on regression but on progression. Peter and Mary's mental journeys give them a "dim sense of some vast, mysterious power, latent in the sub-consciousness of man—unheard of, undreamed of as yet, but linking him with the Infinite and Eternal" (332). This inkling is confirmed after twenty-five years of

---

74. Gustave Le Bon, *The Crowd: A Study of the Popular Mind* (1896; Atlanta: Cherokee, 1982), 34.

75. By the time of Du Maurier's writing, composite photography had already lent itself to representations of biological inheritance. Francis Galton—Charles Darwin's cousin and the originator of eugenics—used it to layer images of faces upon one another, producing single photographs he characterized as pictorial types of hereditary identity, for example of the Jewish "race" (Novak, *Realism, Photography*, 90–93, 95–103). Not coincidentally, probably, Peter Ibbetson speaks of the mammoth he and Mary see as "merely the *type*" of the creature as seen by their prehistoric ancestors (369; emphasis in original).

FIGURE 2. Transported unconsciously to the age of their prehistoric ancestors, Peter and Mary catch sight of a mammoth "blurred and indistinct like a composite photograph." George Du Maurier, *Peter Ibbetson*, 1891.

unconscious togetherness when Mary fails to show up one night in Peter's dreams because, as he later learns, she has died of a heart attack, which pitches him into such a state of insane grief that he must be moved to an asylum. But when he ventures to dream true again, she reappears to impart what she has discovered after death: living beings are contributing to an increasingly more perfect whole and will continue to do so, never reaching a conclusive end. In the hereafter each soul keeps its memories of life, yet also blends as one drop in an ever-accruing cosmic sea in which the good rises to the top and the bad floats to the bottom. Peter plans to write all of the information Mary periodically conveys to him in his dreams but dies before the project is barely begun. The last page of his unfinished memoir contains a sketch of a boy pushing a toy wheelbarrow "from one open door to another. One door is labelled *Passé* [the Past], the other *Avenir* [the Future]" (417).

Peter's mad first reaction to Mary's death, though he tells us that he immediately recovered from it, just leaves open the interpretation that his memoirs are the product of delusion. But that cagey countenancing of spiritual futures may have simply been most appropriate to an era wherein individuals combined a desire for immortality with scientific cautiousness about the proposition.[76] Nonetheless, it is the narrative longing for such a reality that should command our literary historical attention. To ignore it is to misunderstand how many writers were caught up, more or less intently, in the hypothesis of life after death and how even mechanical technologies like the phonograph or camera could become a means of imagining non-mechanical, metaphysical happenings. Du Maurier's novel, for one, reinforces the feasibility of Peter's revelations in having his cousin Madge, who frames his memoirs, affirm, "I, for one, believe him to have been sane, and to have told the truth all through" (6).

Du Maurier's third novel, *The Martian* (1897), essentially refashions Mary's last unconscious visitations and her progressional mystical message in a way that points up their suggestion of extraterrestrial life. The biography of Barty Josselin is told by his best friend, Bob Maurice (who, in another of Du Maurier's nods to narration's service as a mediating technology, tends to a style he calls "telegraphese").[77] The main drama of the novel begins when Barty is saved from a suicide attempt by a fit of sudden unconsciousness. He wakes to find a letter in his own handwriting but authored by an alien named Martia who tells him she has inhabited his body since his childhood. Martia chose to be "suppressed" on her home planet of Mars because she was not fit to reproduce and Martians adhere to a strict policy against bad breeding (380). Now that she has discovered she can communicate with Barty while he sleeps, she instructs him to keep writing materials near his bed at night so that she can dictate for the benefit of those on earth her wisdom about the immortality of the soul, as well as the importance of avoiding marriage and procreation unless one possesses a good physical, mental, and moral constitution. Martia informs Barty that prudent pairings will develop over time individuals' inner "sense of the Deity" (417). Bob Maurice's biography of Barty is meant to honor the man whose multiple books on these subjects, the result of Martia's dictations, have annulled humans' fears of death and raised their average height four to six inches in just thirty years.

Barty records his somnambulistic prophecies of spiritual continuation in the shorthand he learned to write from Bob. Hence Du Maurier's last novel

---

76. Perhaps the epitome of this cageyness is James's *The Turn of the Screw* (1898): as Luckhurst argues, the ambiguity of whether the ghosts are real or only hysterical imaginings is a product of contemporary psychical research's ambivalent response to weird phenomena (*Invention of Telepathy*, 241–42).

77. George Du Maurier, *The Martian* (1897; London: Harper and Brothers, 1898), 3, 24, 94. All further page references to *The Martian* are parenthetical within the text.

replaces the unconscious phonograph with unconscious phonography. The move suggests how easily secretarial writing could be identified with weirder forms of dictated mediation at this time. Just what kind of weirdness Barty is exhibiting remains unclear to him; Martia may be who she says she is, or she may be the spirit of his dead mother or of some other ancestor or, finally, some dissociated part of himself. As a romance the novel rests on the conceit of Martia's validity, but Barty's confusion on the point, which is never quite resolved, reasserts the complex imbrication of psychological and occult explanations for the "ghosts" of the unconscious mind.

The Martian is also plainly invested in Darwinian science and its eugenicist offshoot, though in a manner more complicated than first appears. After years of writing through Barty, Martia decides she would like to fully incarnate herself in his next child. But Marty, the daughter in whom she is reborn (unaware of any prior existence), is alone of the Josselin children in being sickly, and she ends up paralyzed by a spinal injury after she falls out of a tree. Her eventual death from "mere influenza" at the age of seventeen seems most dramatically devastating to Bob Maurice, for whom she has become a darling, and who gradually and unexpectedly becomes the focus of The Martian by its final pages (470). As in Trilby, the narrator (whose surname in The Martian is tellingly similar to the author's) is tacitly offered as a site of the reader's emotional identification. Martia's melioristic evolutionary program is ultimately interrogated by her reincarnation in a degenerated form. If Marty's death is necessary because she is so much frailer than her gorgeous parents, Bob's grief—overwhelming enough that he begs off narrating pivotal parts of her final illness—suggests the high price that comes with the principle of hereditary improvement.

Then there is the tragedy of Bob's own unfitness. We gather that the main reason he has become so close to Marty and a constant avuncular presence in the Josselin family is that Martia's philosophies have convinced him of the immorality of starting his own family. "I'm no Adonis myself," he interrupts his praise of Barty's influential books to note. "I've got a long upper lip and an Irish kink in my nose, inherited perhaps from some maternally ancestral Blake of Derrydown, who may have been a proper blackguard!...Anyhow, I have known my place. I have not perpetuated that kink" (386–87). Our first glance at this commentary takes in its conventional late Victorian linkage of degeneracy with a maligned ethnicity.[78] But noteworthy, too, is Bob's sense

---

78. Edgar Rosenberg was an early critic to note that Du Maurier's Jewish bodies especially are marked as racially inferior, in From Shylock to Svengali: Jewish Stereotypes in English Fiction (Stanford: Stanford University Press, 1960), chap. 10. Compare Neil R. Davison's more recent "'The Jew'

of obligation or "place" as a bachelor and, concomitantly, the novel's unusual exploration of the painful personal implications of late Victorian theories of biological advancement. At the time of Bob's writing, Barty and his wife, Leah, are both dead and their children grown, and Bob's feelings of isolation are so profound that they inevitably bubble up into his narrative: "And now, gentle reader, I want very badly to talk about myself a little, if you don't mind...I feel uncommonly sad, and very lonely indeed" (422).

Given the Victorian feminization of automatism, it would seem odd that Martia's conduit to humanity is male, and Peter Ibbetson is another man liable to subliminal transmissions. Importantly, though, each hero's adventures entail as a key feature his conscious processing or reprocessing of unconscious knowledge, in a way that underscores his own identity. Peter's true dreams basically embed a conscious state within an unconscious one: "my consciousness, my sense of being I, myself," usually in dreams "partial, intermittent, and vague, suddenly blazed into full, consistent, practical activity—just as it is in life, when one is well awake and much interested in what is going on—only with perceptions far keener and more alert" (202). As for Barty, he is responsible for reworking by day the notes Martia transmits nightly by shorthand; what is even more significant, his automatic writings ultimately bear the stamp of personal authorial claim, bringing literary fame to his own name.

Further, both *Peter Ibbetson* and *The Martian* contain emblems of conventionally gendered mediation. It is Mary who shows Peter how to dream true, approximating the séance medium's task of bringing forth to him the "beloved shades" of departed friends and family (274). Later she fulfills the medium's function more completely in channeling wisdom from immortal regions. In *The Martian,* as an aid to his writing, Barty's wife, Leah, takes on the "happy labour of love" of acting as his amanuensis (363).[79] However, each of these women, Mary and Leah, is also plainly intelligently self-possessed. Bob remarks about Leah, for example, that she is not "the kind of woman that makes a door-mat of herself for the man she loves" (409), and she openly scoffs at Barty's more pretentious phrasings and insists he change them. All this contrasts with La Svengali's complete vacuity and amnesia as her maestro's automaton. She never does discover how he has used her. One may wonder if some part of the reason *Trilby* made more of a splash than Du Maurier's other two novels is that it corroborated stock notions of how hypnotic

---

as Homme/Femme-Fatale: Jewish (Art)ifice, *Trilby,* and Dreyfus," *Jewish Social Studies* 8 (2002): 73–111.

79. As his blindness progressed, Du Maurier's own wife, like his children, had served in the same capacity during his writing; Ormond, *George Du Maurier,* 442.

somnambulism and other kinds of unconscious transmission work—that is, of men and women's relative roles in these operations. When Victorian narratives assign mediation, especially occult mediation, to men, they often imply the medium's effeminacy, as my final chapter shows.

What *Peter Ibbetson* and *The Martian* do have in common with *Trilby* is a keen sense of the powers of the unconscious mind, especially its capacity to channel strange and wonderful communications. Du Maurier's media produce occult effects that imply the transferability of the self and the impressionability of the sub-psyche. Moreover, they allow us a glimpse of a world yet to be fully realized, where the self combines with other selves in vigorous new formations, from the modern popular crowd to the cosmic soup of the "Infinite and Eternal." The recording unconscious of Du Maurier's somnambulists lets out onto the future, enabling thrillingly unfamiliar ranges of communion and experience.

❧ CHAPTER 4

Telltale Typing, Hysterical Channeling

The Medium as Detective Device

The term *occult* makes a statement about knowledge. It is at once a barrier and an invitation, demarcating the hidden even while challenging us to make it known. This is the same sort of dual meaning that governed modern spiritualists' understanding of their enterprise. Claiming that the séance worked by as yet imperfectly understood natural laws, they denied that the question was what could *be*—spirits were indubitably real—and recast it as a matter of what could be *known*. For many, indeed, spiritualism turned on a sense of mystery and its opposite. Reading séance accounts, one is struck by the fact that participants so often tolerated the medium's curtained partition as a precursor to ghostly phenomena. Given that the situation was so vulnerable to fraud, why would séance sitters allow the medium a space free from witnesses? Perhaps precisely because the curtain thwarted observation, it stood for the occult itself and became naturalized as part of the experience. The machinery of having to pull back a screen was indispensable insofar as it dramatized the revelatory essence of the séance.

If the séance turned on a dynamic of ignorance and knowledge, it made women into detective devices: instruments for probing and discovering. What mediums divulged could be existential data or, more simply, mortal secrets. At the birth of the modern spiritualist movement, the raps the Fox sisters inspired in their home supposedly originated from the ghost of a peddler furtively murdered and buried in their cellar. Other spiritualistic secrets were ones of the heart, as in the strange case of the "cross-correspondences." Beginning in 1901 and continuing for decades, four mediums generated thousands of automatic scripts based on a unified sign system—a telegraphic array of symbolic words, images, and literary allusions corresponding to a single semantic code—despite the fact that the four women had each done their transcriptions independently and were not all known to one another. The scripts concerned a confidential love affair the mediums had been wholly unaware of, between one Mary Lyttelton, who had died in the 1870s, and the still-living politician and psychical researcher Arthur Balfour.[1] As Lyttelton's mediated professions of enduring love demonstrate, the spirit channel transgressed a threshold not just between ignorance and knowledge but also between privacy and publicity, and in that respect, she was not unlike other female media. Typists and operators readied words meant for specific recipients to travel away from their senders across sometimes broad expanses. As such they tended a blurry line delimiting private conversation and knowledge.[2] One way to think about the séance is indeed as a distillation of this increasingly common circumstance of modern technologies. Compressing time and space into one room, the séance was a hyperbole for the social reach of modern communications. Remoteness bred a networking that made information more accessible, with the relay being the primary point of accessibility.

In calling female relays detective devices, I aim to shed a brighter light on instances we have already glimpsed in which women serve as tools for discovery and as potential nodes for publicity. My point in this chapter is to see how these women figure in narratives more strictly defined by their revelatory plots. It is probably little coincidence that the typist appears repeatedly in detective fiction, with its genre-defining emphasis on the unveiling of illicit secrets. On the face of things, the rhetoric of her mechanism makes her little more intelligent than the typewriter itself, an apparatus that in Arthur Conan Doyle's "A Case of Identity" (1891) becomes crucial to solving the mystery. Other detective tales, however, depict the mechanical typist in ways that allow her a more substantive part in the hunt for information; these reinforce that

---

1. On the cross-correspondences, see Jean Balfour, "The 'Palm Sunday' Case: New Light on an Old Love Story," *Proceedings of the Society for Psychical Research* 52 (1960): 79–267, and Janet Oppenheim, *The Other World: Spiritualism and Psychical Research in England, 1850–1914* (Cambridge: Cambridge University Press, 1985), 132–35.

2. As Jonathan Sterne notes, the telephone, like the telegraph, "facilitated intimately personal connection *because* it was a massive network of connections. It was simultaneously intensely public and intensely private." *The Audible Past: Cultural Origins of Sound Reproduction* (Durham: Duke University Press, 2003), 208.

the idea of female automatism ultimately opened up multiple responses to the question of the place women could occupy within knowledge networks.

Psychologist Morton Prince's popular case study *The Dissociation of a Personality* (1905) offers a somewhat less vexed rendering of his patient's unconsciousness. A victim of nervous disorder, Christine L. Beauchamp (an alias for Clara N. Fowler) represents yet another version of the remarkably sensitive, apparently unassuming female relay, in that Prince's treatment revolves around his dealings with the alternate personalities her body channels. His hypnotism aids him in unearthing a fully realized subconscious self, the fetching star of the case. One inevitably wonders to what degree Prince's patient really was unassuming and what she may have hoped to gain in exhibiting the illness she did. But my analysis here, unlike that of the typists in the work of Doyle and others, does not dwell on the medium's personal investment in the project of discovery. Rather, I am interested in how Prince's own narrative of this project is indebted to a then-popular discourse, that around non-mortal possession. Certain flourishes deliberately remind the reader that the selves he brings into public view are only the last in a long line of occult presences. In the end, though, such stylings, along with his treatment, both sensationalize Prince's premise of detection and trouble it, suggesting an uncertain relationship to what he calls knowledge.

Typing over Irene Adler: "A Case of Identity"

In Arthur Conan Doyle's "A Case of Identity," Sherlock Holmes pins down his antagonist by tracing him back to his typewriter. Like all typewriters, Holmes says, this one has acquired several imperfections of marking that give it "really quite as much individuality as a man's handwriting," making it possible to match up letters generated on it to the man who has written them.[3] Yet the actual circumstances of the writing remain obscure. Did the villain, James Windibank, type these letters himself? Or—since as he tells Holmes the machine is one used for correspondence by his whole office—did a clerical worker type them for him? The latter scenario initially seems less likely, given the intimate nature of this particular correspondence. Windibank has been writing love letters to his stepdaughter, pretending to woo her in the guise of a fictional and now vanished persona, Hosmer Angel, to whom

---

3. Arthur Conan Doyle, "A Case of Identity," in *Sherlock Holmes: The Major Stories with Contemporary Critical Essays,* ed. John A. Hodgson (Boston: Bedford, 1994), 86. All further page reference to "A Case of Identity" are parenthetical within the text.

he knows she will remain faithful, avoiding marriage and thus leaving her income for his use. Yet this scenario is also curiously gestured toward in the profession of the stepdaughter herself, Mary Sutherland, a typist.

When Miss Sutherland seeks out Holmes to find the missing Hosmer Angel, she tells him about her own love letters: "I offered to typewrite them, like he did his, but he wouldn't have that, for he said that when I wrote them they seemed to come from me but when they were typewritten he always felt that the machine had come between us" (80). Notably, Hosmer's request that she handwrite her letters is never explained as part of Windibank's ruse. It is therefore allowed to stand as figuratively significant of typing's different capacities for expression between the two sexes. Both within and without the fiction of Hosmer's existence, typing is tied to the individual self of the male writer. On the one hand, it is good enough for Hosmer to communicate his heartfelt emotions to his lover; on the other hand, it is, according to Holmes, individual as a *"man's* handwriting" and therefore capable of giving away Windibank's identity. But a woman's association with the typewriter is quite another matter. When Miss Sutherland types, she is usually communicating someone else's thoughts and will, not her own—which would explain why the instrument cannot, in her fiancé's estimation, adequately convey herself. Her position at the typewriter undercuts her status as both a singular and thinking being. In fact, as described in the rest of Doyle's story, the female typist is so oblivious a body as not only to transcribe without obtruding upon the most private of communications, for example Windibank's fake love letters, but also to be, psychologically, virtually continuous with the kind of machine that reveals his deception.

"A Case of Identity" posits this specific linking of women, technology, and issues of intelligence in part by implicitly contrasting it with another constellation of these terms, one suggested in a Holmes story published only a few months earlier, "A Scandal in Bohemia" (1891). The latter gives us a character notorious among Sherlock Holmes aficionados for having outsmarted him, a feat all the more remarkable because accomplished by a woman. In "Scandal" the King of Bohemia hires Holmes to recover a photograph of himself with his ex-lover, Irene Adler, before she can send it to his scrupulous fiancée. Holmes uses a disguise to enter her house to discover the photograph's hiding-place and thinks afterward that the plan has worked, yet the next day turns "white with chagrin and surprise" to find Irene has fled the country, taking the photograph with her.[4] As she describes in a letter she

---

4. Arthur Conan Doyle, "A Scandal in Bohemia," in *Sherlock Holmes: The Major Stories with Contemporary Critical Essays,* ed. John A. Hodgson (Boston: Bedford, 1994), 50. All further page references to "A Scandal in Bohemia" are parenthetical within the text.

leaves behind, she saw through his disguise and even donned one of her own to follow him back to Baker Street to verify her suspicions.

When the King, grateful she at least no longer plans to impede his marriage, offers Holmes his emerald ring as reward, the detective responds, unexpectedly, by asking instead for a photograph of herself that Irene has enclosed for the King with her letter. The King has an elitist's blind spot: "What a woman—oh, what a woman!...Would she not have made an admirable queen? Is it not a pity she was not on my level?" But Holmes knows better, as his "coldly" sarcastic retort shows: "From what I have seen of the lady, she seems, indeed, to be on a very different level to Your Majesty" (52). In parting, Holmes "turn[s] away without observing the hand which the King had stretched out to him," apparently offended that the King has undervalued so formidable a companion. Watson's final words indicate how chastened his friend feels at having been "beaten by a woman's wit. He used to make merry over the cleverness of women, but I have not heard him do it of late. And when he speaks of Irene Adler, or when he refers to her photograph, it is always under the honourable title of *the* woman" (52).

Here as in the Mary Sutherland case, technology pegs identity. But while this works against James Windibank, it works for Irene Adler, because a photograph visually proves her part in the King's life. And by the conclusion of the story photography has come to symbolize something more, Irene's superior intellect—not once but twice over, in the form of her two photographs. The photograph threatening to the King finally eludes the detective, obviously marking his failure. Irene's other photograph, which ends up in Holmes's possession, functions for him as a keepsake attesting to her success and, by extension, to his previous gender misconceptions. Her "honourable title of *the* woman" in his vernacular suggests his sudden respect for her as his intellectual equal. Talking about either her or her photograph, Watson says, reminds Holmes that he erred, and the error has everything to do with women's "wit" and "cleverness."

But "A Case of Identity" substitutes Irene Adler's wit vis-à-vis one technology with Mary Sutherland's witlessness vis-à-vis another. This thematic erasure is signaled at the very outset of "Identity" by its revisionist perspective on certain plot details in "Scandal." In an only seemingly throwaway exchange in "Identity," Watson remarks on a rich snuffbox Holmes extends to him; Holmes replies, "I forgot that I had not seen you for some weeks. It is a little souvenir from the King of Bohemia in return for my assistance in the case of the Irene Adler papers" (76). Significantly, this response elides Irene's photographic self-testimony, and with it Holmes's hard lesson about women, in two respects. First, whereas "Scandal" ended with Holmes's dramatic request for Irene's photograph, his reply here effectively replaces this with another "souvenir" of the case. Moreover, this souvenir attests to the power of the wealthy King instead of any Irene has proven by her triumph. Indeed, Holmes has made a total about-face: he declined the King's emerald ring before, but has now accepted this gift of a snuffbox.

Second, Holmes now inexplicably refers to the item coveted by the King as "papers." This inaccurate reference is not trivial, because it retrospectively expunges the fundamental source of his adversary's pull. For as Holmes explicitly stated in "Scandal," nothing, certainly no papers, could intimidate as much as the photograph in Irene's possession. Holmes queries the King upon first meeting him:

> "I fail to follow Your Majesty. If this young person should produce her letters for blackmailing or other purposes, how is she to prove their authenticity?"
> "There is the writing."
> "Pooh, pooh! Forgery."
> "My private note-paper."
> "Stolen."
> "My own seal."
> "Imitated."
> "My photograph."
> "Bought."
> "We were both in the photograph."
> "Oh, dear! That is very bad! Your Majesty has indeed committed an indiscretion." (39)

Only a photograph, with its seemingly indisputable visual record, holds the secret of this intimate relationship so damning to the King. Even when the facts of the case are recalled more plainly later in "Identity," this occurs only in the context of one of Watson's paeans to Holmes's genius: "I had had so many reasons to believe in my friend's subtle powers of reasoning, and extraordinary energy in action...Only once had I known him to fail, in the case of the King of Bohemia and of the Irene Adler photograph, but when I looked back to the weird business of the Sign of Four, and the extraordinary circumstances connected with the Study in Scarlet, I felt that it would be a strange tangle indeed which he could not unravel" (85).

Holmes's own refashioning of crucial moments around the Bohemian affair forecasts a return in "A Case of Identity" to a conservative ideology of relative mental abilities between the sexes, one that forcefully corrects the personal and gender disruption connoted in his defeat. Specifically, the story's suppression of both of Irene's photographs paves the way for it to deal with another technology with very different implications for ideas of women's brain power. In "A Scandal in Bohemia," Irene Adler uses a photograph to protect her own account of the past against whatever account the King may choose to fabricate. In addition, she has changed the meaning of the photograph itself, transforming it from what we presume was a romantic memento—a picture of some physical or emotional intimacy—into a piece of accusatory evidence. In these ways photography allows her to author, and hold in reserve, her own narrative of the truth. By contrast, "A Case of Identity" implies that the woman at the typewriter lacks authorial importance in the production of facts or ideas and becomes, rather, a mere appendage of the machine.

Holmes and Watson initially glimpse Mary Sutherland standing on the pavement beneath Holmes's lodgings: "she peeped up in a nervous, hesitating fashion at our windows, while her body oscillated backwards and forwards, and her fingers fidgeted with her glove buttons" (76). Her repetitive movements coupled with her mobile fingers identify her with the device it is her job to operate. Holmes's interpretation of the scene further reduces her to a generic, mechanically predictable body: "I have seen these symptoms before…Oscillation upon the pavement always means an *affaire du coeur.* She would like advice, but is not sure that the matter is not too delicate for communication. And yet even here we may discriminate. When a woman has been seriously wronged by a man she no longer oscillates, and the usual symptom is a broken bell wire" (77).[5] The hypothetical "broken bell wire" may indicate an insistent appeal for retribution at the bell outside Holmes's lodgings, which Miss Sutherland has just rung; yet in the context of the rest of his speech, it seems instead to describe the heartbreak of betrayal in terms of the female biological apparatus itself.

Holmes's assumption that women's behavior is so reflexive and formulaic is exactly what trips him up in "A Scandal in Bohemia." In that story he starts a small fire in Irene Adler's home, believing he knows exactly how she will react: "When a woman thinks that her house is on fire, her instinct is at once to rush to the thing which she values most.… A married woman grabs at her baby—an unmarried woman reaches for her jewel box. Now it was

---

5. Arguing for the detective's scientific understanding of identity as materiality, Ronald R. Thomas likewise recognizes that Mary Sutherland's bodily movements mark her as similar to a machine and notes that her bad eyesight and touch typing result in her "automatic" textual production; *Detective Fiction and the Rise of Forensic Science* (Cambridge: Cambridge University Press, 1999), 82. He eventually connects the story to the spate of late Victorian female typists who participated in business communications but in a sightless and therefore "alienated and unconscious form" (87).

clear to me that our lady of to-day had nothing in the house more precious to her than what we are in quest of" (49). Irene does inadvertently reveal the location of the photograph, but Holmes has not counted on her ability to recognize her mistake and to take compensatory measures. His estimation of the automatic regularity of the female mind appears justified, though, in the case of Mary Sutherland, whose face Watson twice describes as "vacuous" (77, 82) and whom one reader has ceremoniously dubbed the "Stupidest Client" in the Sherlock Holmes canon.[6]

Miss Sutherland's vacuity is reinforced through a Holmesian motif of vision as a mode of perceptiveness, tweaked to reflect the age of the typewriter. Holmes's first words to her concern her eyesight: "Do you not find...that with your short sight it is a little trying to do so much typewriting?" "I did at first," she says, "but now I know where the letters are without looking." The exchange immediately opposes Holmes's quick deductive observations—he has gleaned she is a typist without her saying so just by having "looked over her in the minute and yet abstracted fashion which was peculiar to him"—to her myopia, which is in turn exaggerated in her "blind" typing, a common turn-of-the-century method that disciplined the typist's nerves and muscles, making her work seem more automatic.[7] Of course Holmes's claim to fame is, as he says, his ability "to see what others overlook" (77). In this context Miss Sutherland becomes, like the letters from Windibank's typewriter, or the typewriter itself, evidence to be read. "You appeared to read a good deal upon her which was quite invisible to me," Watson says (83).

In the end, Holmes chooses to leave Mary Sutherland in the dark about Hosmer Angel's identity, despite the fact that she is his client and has expressly come to him for enlightenment. "If I tell her she will not believe me," he explains to Watson. "You may remember the old Persian saying, 'There is danger for him who taketh the tiger cub, and danger also for whoso snatches a delusion from a woman'" (89). Notwithstanding Irene Adler, Holmes is back to "merry" truisms that deny women the possibility of complex or spontaneous thought and instead put their behavior on a par with animal instinct.

---

6. Roger Butters, *First Person Singular: A Review of the Life and Work of Mr. Sherlock Holmes, the World's First Consulting Detective, and His Friend and Colleague, Dr. John H. Watson* (New York: Vantage, 1984), 136.

7. See Christopher Keep, "Blinded by the Type: Gender and Information Technology at the Turn of the Century," *Nineteenth-Century Contexts* 23 (2001): 149–73. Keep mentions "A Case of Identity" in a footnote (170 n. 12), but without investigating how Mary Sutherland bears out his own conclusions about gender and "blind" typing. On vision and Mary Sutherland's typing, see also my note 5.

Like the tiger protecting its cub, or the human mother her baby during a fire, Mary Sutherland is ruled by an unthinking familial devotion, in this case her "simple faith" in her promised husband, Hosmer Angel (82).

Detective Mechanisms of the Girl behind the Keys

Tom Gallon's *The Girl Behind the Keys* (1903) revolves around the same conception of automatic female behavior that makes Mary Sutherland a synecdoche for the typewriter, but with a crucial twist on the meaning of this behavior. The novel follows the adventures of Bella Thorn after she is hired on as secretary to what turns out to be a criminal group. As Pamela Thurschwell points out, Bella acts naive and harmless, answering any concern of her boss, Neal Larrard, about the safety of his nefarious secrets by telling him that as a typist she operates mechanically, even as she stealthily works to frustrate his plots.[8] What I am interested in is how secretarial unmindfulness becomes, then, a deliberate pretense and, moreover, a way not just of denying knowledge but at the same time of proactively gaining it—in other words, of becoming a detective. *The Girl Behind the Keys* turns into a detective story when Bella's journalist boyfriend asks her to stay on with Larrard's gang to find out more about them and help bring them to justice. The newspaper report of the trial that closes the book describes her as a witness who had "closely...studied the methods of the gang."[9] Yet Larrard's hiring of Bella conveys the extraordinary presumption that a secretary poses little threat to the confidentiality of even the most egregious of businesses. He stipulates during her interview, "we deal only with special clients. Discretion is necessary; that is to say, if a secretary is sent out to one of these clients, she must forget, at the end of a day, all that happened during that day" (8). By its very tenacity, the trope of the "forgetful" office worker serves as an unexpected means to female detection.

---

8. Pamela Thurschwell, "Supple Minds and Automatic Hands: Secretarial Agency in Early Twentieth-Century Literature," *Forum for Modern Language Studies* 37 (2001): 155–68. Thurschwell sees in Bella's story an example of an early twentieth-century division in perceptions of clerical functions between the simply mechanical typist-stenographer and the more esteemed secretary, who has the necessary manual skills but also "identifies and sympathises" with the interests of her employer, as a wife would (156). Thurschwell views the contemporary telephone operator's "civil-listening" and pacifying as similarly demonstrating a combination of "mechanical and comforting" behavior and as similarly based in domestic priorities (159). The secretary's wifely "eroticisation" in relation to her boss also means she becomes a "protector of intimate secrets" of his work (164).

9. Tom Gallon, *The Girl Behind the Keys* (London: Hutchinson, 1903), 125. All further text references to *The Girl Behind the Keys* are parenthetical within the text.

Bella can in fact sometimes be rather gullible, lagging behind the reader in understanding Larrard's schemes. Yet she always discovers the truth and is quicker to glean it by the later chapters, so that in the final estimation Gallon seems largely to be developing in order to better explore the possibilities for knowledge latent within the canard of the clueless mediating woman. In this light Bella's repeated comparisons of herself to a machine are equivocal, on the one hand sincerely describing her bodily efficiency but on the other highlighting the irony of her position, the mere semblance of her unthreatening proximity to covert misdeeds. Whatever anxieties about her presence Larrard may have are easily allayed for a while: "Having always a placid exterior, I was, to all appearances, as much a machine as that at which I worked; and I think, in time, he began to cease to think that I need be considered at all" (35). Even when he grows more concerned that she may not really "forget," her position still allows her to play the obedient servant as she monitors and plans rescues of his would-be victims: "through it all, I had to keep a calm, quiet face, and go about my work like a machine" (94). Bella's typewriter provides her a front, as Gallon's title cleverly underscores. Like Irene Adler, she adopts the famous Sherlock Holmes's strategy of disguise, but with the disguise here being notional rather than physical, constituted by an ideal image of herself as medium—by the gap produced in every stereotype between what is supposed to be and what actually is.

A few decades later, Dorothy Sayers would likewise represent women's mediation as a façade for their detection. In *Strong Poison* (1930), Lord Peter Wimsey finances a faux typing bureau whose workers serve as his under-sleuths and which, in an apparent nod to their feminine slyness, he calls "My Cattery." One of these typists, hired on at a suspicious solicitor's office, executes her mission by playing the absent-minded worker, purposely mistyping a document; having to correct her work gives her an excuse for staying after hours and snooping. Her coworker, who "regret[s] the intrusion of female clerks" into his profession, takes the "careless" product of her and her machine as a sign of the changing times.[10]

Wimsey next dispatches the Cattery's head, the appropriately named Katharine "Kitty" Climpson, to find an elusive will. Her sex and the moral presumptions around it place certain constraints upon her when she attempts to follow the testatrix's nurse: "The male detective . . . is favourably placed for 'shadowing.' He can loaf without attracting attention. The female detective must not loaf. On the other hand, she can stare into shop-windows for ever.

---

10. Dorothy Sayers, *Strong Poison* (1930; New York: HarperCollins, 1995), 155, 157.

Miss Climpson selected a hat-shop."[11] In other words, the trick, for her as for her colleague, is to make detection look like feminine distraction. Accordingly, in the hilarious chapters that ensue, Miss Climpson ingratiates herself with the spiritualistic nurse by feigning a gift for automatic writing. Like the caper in Gallon's work wherein Larrard uses Bella to send ersatz spirit messages through a rigged typewriter, Miss Climpson's ploy also suggests that, at a basic level, the transmissions of the typist are conceptually of a piece with those of the séance medium.[12] Whole pages are devoted to the cagey alphabetic raps and furtively penciled-out Ouija-board phrases through which she gathers clues. Along the way, Miss Climpson exploits the principle of the medium's feeling receptivity. "I've been told I'm sensitive myself, but I have never dared to *try*," she says upon her first meeting with the nurse, who eagerly takes the bait: "You are such a sympathetic person. I'm sure we should get good results."[13]

Detective fiction trades on secrets, and as it turns out, the mediating woman within it sometimes harbors a secret of her own. In *The Girl Behind the Keys* and *Strong Poison,* she absorbs the difference between the attributes we see in the two Holmes stories—feminine intelligence and feminine stupidity—into her own character, using that complicated knowledge position to conceal her aims of abetting others' factual quests. The cliché around women's typing opens up a space for a productive duplicity that silently queries the cliché itself.

The Haunting of Miss Beauchamp

In the opening scene of *The Sign of Four* (1890), Watson tries to test what he thinks is his friend's overstated claim to deductive ability. "I have heard you say it is difficult for a man to have any object in daily use without leaving the impress of his individuality upon it in such a way that a trained observer might read it," Watson says, handing Holmes a watch, smugly challenging him to detect from this slim evidence the "character or habits of the late owner."[14] Holmes greatly exceeds Watson's expectations, of course, rattling

---

11. Ibid., 183.

12. Thanks to Clare Simmons for pointing my way to *Strong Poison*'s depiction of a typist-cum-spirit medium.

13. Sayers, *Strong Poison,* 188, 189.

14. Arthur Conan Doyle, *The Sign of Four,* intro. Peter Ackroyd, with notes by Ed Glinert (1890; London: Penguin Books, 2001), 9.

off a brief history of the watch's owner, demonstrating the height of his quasi-scientific perception. At the same time, Holmes's divination of the watch owner's "impress of his individuality" suggests a different mode of perception, one professed by certain nineteenth-century psychics: psychometry, in which one senses the personal background of an object merely by touching it. Holmes's deduction has little to do with the paranormal. Still, it is not hard to imagine that Doyle, who became a member of the Society for Psychical Research in the 1890s and later published books on spiritualism, would have noted similarities between Holmes's attentiveness and that of the psychometrist: in both cases objects render up occult meanings to an agent acutely sensitive to their physical characteristics.

As I have suggested, Victorian occultists were always more or less engaged, like Sherlock Holmes, in detection—of unspoken thoughts, faraway scenes, spirits, the afterlife. By the end of the nineteenth century, another mode of detection had also gained prominence, one that seemed to have little to do with the paranormal and in fact recast in mundane terms the effusions of the séance. If a medium channeled strange words or thoughts, some proposed, these came not from some other world but rather from her own mind. In certain instances, that is, psychical researchers like Frederic Myers pointed as another explanation to the existence of unconscious trains of consciousness beneath our primary selves, as did psychologists.[15] The idea excited much interest within discussions of the hypnotic trance—often viewed as a portal to the unconscious—and especially of double consciousness, a condition in which unconscious selves competed with the primary self, wresting control of body or speech. Patients switched from one consciousness to another in certain widely reputed cases that anticipate modern accounts of multiple personality (dissociative identity) disorder. Since the beginning of the diagnosis, a majority—by some calculations a very large one—of such cases have involved female patients.[16] The hysteric acted as middlewoman between her unconscious self or selves and the hypnotist wanting to learn more about them.

Along with Boris Sidis, psychologist Morton Prince, who authored a lengthy case study of multiple selfhood titled *The Dissociation of a Personality*, was a top voice on abnormal psychology in early twentieth-century America.[17] But *Dissociation* is not merely targeted to other psychologists. Prince later

---

15. On Myers's views of the possible confusion between spirits and subliminal expressions, see Adam Crabtree, *From Mesmer to Freud: Magnetic Sleep and the Roots of Psychological Healing* (New Haven: Yale University Press, 1993), 331–32, 342.

16. Ian Hacking, *Rewriting the Soul: Multiple Personality and the Sciences of Memory* (Princeton: Princeton University Press, 1995), 69.

17. On Prince's professional status, see Robert C. Fuller, *Americans and the Unconscious* (Oxford: Oxford University Press, 1986), 105.

admitted to enlarging upon the case's most dramatic moments, and his book created much public interest in the condition of dissociated consciousness and was reviewed in several popular periodicals, even becoming the basis of a play.[18] The first chapter promises amusement ("amusing" is surely one of the book's most frequent non-clinical adjectives), describing the vicissitudes of his patient's condition as material for "a multitude of plots for the dramatist or sensational novelist."[19]

Insofar as such moments hint at Prince's desire to entertain, it makes sense to focus on his work as a popular literary one. He implies drama and the sensation novel as influences, but his book also recalls another genre, a certain turn-of-the-century breed of narrated detection involving the discovery of other- or netherworldly presences.[20] Hypnosis is a primary tool in the case, and in *Dissociation,* as in other narratives, the hypnotic scene lends itself to the coming together of scientific with occult interests. Further, like many fictional and nonfictional texts of the time period, Prince's writing demonstrates the fluidity between psychical research and psychology due to both fields' interest in the unconscious and hypnosis. *Dissociation* alludes to writings by psychical researchers Frederic Myers and Edmund Gurney and contains citations to the *Proceedings of the Society for Psychical Research,* which had carried an early report of his case.[21] But whereas we might expect that in a study of this sort Prince would distance himself from the weirder interests of the SPR, instead he stylistically amplifies the gray area, representing his patient as magically invaded. Though such moments are scattered and far from terrifying, they are not accidental—not merely the product of convenient or shared images or methods, as for example is the case in Sidis's work—but instead opportunistic nods to supernaturalist literature and ways of tapping into public curiosity about psychical research.[22] Noticing such shadings helps us begin to discern Prince's other important effects as a creative writer: his clearly—indeed, artificially—defined characters, along with the exaggerated struggles over gender

18. Michael Kenny, *The Passion of Ansel Bourne: Multiple Personality in American Culture* (Washington, DC: Smithsonian Institution Press, 1986), 139–40, 156–57.

19. Morton Prince, *The Dissociation of a Personality: A Biographical Study in Abnormal Psychology* (1905; New York: Longmans, Green, 1906), 2. All further page references to *The Dissociation of a Personality* are parenthetical within the text.

20. For Ruth Leys, the case resembles not only the detective story (traditionally conceived) but also the adventure story and the fairy tale. *Trauma: A Genealogy* (Chicago: University of Chicago Press, 2000), 42.

21. Morton Prince, "The Development and Genealogy of the Misses Beauchamp: A Preliminary Report of a Case of Multiple Personality," *Proceedings of the Society for Psychical Research* 15 (1901): 466–83.

22. Prince also draws on literature in his representation of psychical duality, as Karl Miller elaborates in *Doubles: Studies in Literary History* (Oxford: Oxford University Press, 1985), 334–38.

identity that the clashes between these characters imply. Moreover, we see how hypnosis allows him other opportunities for exercising a creative will, one that is ultimately at odds with his general claims of discovery.

It is worth giving a detailed summary of Prince's case for the issues it raises. When "Christine Beauchamp" comes to him for treatment in 1898, he diagnoses her as a neurasthenic (a sufferer of nervous exhaustion) and uses hypnosis to relieve her symptoms. But the treatment takes a startling turn when, while entranced, she begins to exhibit a personality different from both her waking and usual hypnotic states, which claims to have existed throughout Miss Beauchamp's life, aware of all Miss Beauchamp's thoughts and actions yet unable to surface physically and accomplish her own aims. Soon Sally, as Prince names this second personality, figures out how to assume bodily control and alternate regularly with the first personality, who is highly religious and reserved and whom Sally, careless and roguish, sees as a killjoy and enjoys tormenting. Prince distinguishes his objects of study as BI (the self who sought therapy), BII (the hypnotic self), and BIII (Sally). Then in 1899, another personality comes to light; this one, called BIV, is amnesiac for the previous seven years of BI's life.

Bent on discovering the "Real Miss Beauchamp," and believing she must be emotionally and nervously resilient, he is inclined to think the collected, self-confident BIV fits the bill and has merely been supplanted by BI, until BIV too begins to evince an abnormal sensitivity to her environment. BI and BIV are each oblivious to the other's experiences; Sally alone fits Prince's definition of a subconscious or co-conscious personality because she exists continuously beneath the consciousnesses of all the others. Prince theorizes that the Real Miss Beauchamp will not only be healthy and stable but also fuse the personalities of BI and BIV, which he concludes were dissociated seven years ago due to an emotionally shocking event that involved some murkily sketched confrontation with a family friend named Jones. Because BII represents just such an amalgam, he eventually suspects that she is the Real Miss Beauchamp, somehow trapped in a hypnotic state. After some botched attempts due to Sally's interference, he manages to lift BII out of hypnosis, intact with the thoughts and memories of both BI and BIV. As for Sally, she ends up, as she calls it, "squeezed" out of active existence, consigned to her former muteness (405).

In many respects the pathologically nervous woman stepped easily into the conceptual framework occupied by the spirit medium. The sensitive machine was the same, only labeled differently. As Jeffrey Sconce suggests, even while dismissing spiritualists' claims that the medium's "surplus (or imbalance) of 'nervous energy' made her a more receptive candidate for receiving the

higher electromagnetic transmissions of the spirits," nineteenth-century doctors' diagnoses of neurasthenia and hysteria entailed comparable views: "This
surplus led instead to dysfunctions of the body, where the nervous system, as a
great telegraphic network, was overtaxed by the variable intensity of this flow."[23]
Like the doctors Sconce discusses, Prince first responds to his patient's neurasthenia with electrotherapy, made famous by men like George Beard, who
imagined the nervous system as an electric generator whose supply could be
artificially stimulated.[24] Electromagnetism continues to serve in Prince's therapy once he turns to hypnosis.[25] In this respect he demonstrates a trend in
turn-of-the-century hypnotic treatments. The students of eminent hypnotist
Jean-Martin Charcot effectively revived the tenets and practices of Mesmer,
applying magnets to control the emotions of their hysterical patients.[26] Similarly, William James's "The Hidden Self," a 1907 essay in *Scribner's Magazine*,
tells of hypnotists employing metals, magnets, and electrodes, as when Pierre
Janet "restored... tactile sense temporarily by means of electric currents, passes,
etc."[27] Prince himself mentions that he sometimes uses "a strong Faradic battery to wake Miss Beauchamp when [Sally] would not go at command" (96).

Prince plays up the resemblance between hysterical and occult communication to generate suspense about the hidden beings, especially Sally, revealed
by his patient. While he nicknames BI "The Saint" and BIV "The Woman,"
Sally is "The Devil," and at one point he recounts a rather histrionic mock
ceremony: "Putting my finger to her forehead, I made her believe I had the
power of exorcism. The effect was remarkable. She shrank from me much
as the conventional Mephistopheles of the stage shrinks from the cross on
the handle of the sword" (137).[28] But as he clarifies, Sally's demonism is more

---

23. Jeffrey Sconce, *Haunted Media: Electronic Presence from Telegraphy to Television* (Durham: Duke
University Press, 2000), 51–52.

24. On electrical treatments of neurasthenia, see Tim Armstrong, *Modernism, Technology, and the
Body: A Cultural Study* (Cambridge: Cambridge University Press, 1998), chap. 1; and John S. Haller,
Jr., and Robin M. Haller, *The Physician and Sexuality in Victorian America* (Urbana: University of Il-
linois Press, 1974), 5–23.

25. Saul Rosenzweig notes Prince's use of electricity in "Sally Beauchamp's Career: A Psychoarchaeological Key to Morton Prince's Classic Case of Multiple Personality," *Genetic, Social, and General
Psychology Monographs* 113, no. 1 (1987): 51.

26. Anne Harrington, "Hysteria, Hypnosis, and the Lure of the Invisible: The Rise of Neo-
Mesmerism in *Fin de Siècle* French Psychiatry," in *The Asylum and Its Psychiatry*, ed. W. F. Bynum, Roy
Porter, and Michael Shepherd, *Anatomy of Madness: Essays in the History of Psychiatry*, vol. 3 (London:
Routledge, 1988), 226–46.

27. William James, "The Hidden Self," *Scribner's Magazine* 7 (1890): 366.

28. Analyzing this scene and others, Kenny is attuned to the ideas of possession as well as literary
precedents informing Prince's case (*Passion of Ansel Bourne*, 146); however, he sees these contributions
as unintentional rather than, as I read them, crucial to Prince's design of character and narrative
conflict.

playful than anything else: she is "not an immoral devil, to be sure, but rather a mischievous imp, one of that kind which we might imagine would take pleasure in thwarting the aspirations of humanity" (17). Light-heartedly naughty, Sally resembles the untoward spirits that sometimes thwarted séance sitters' aspirations of meaningful exchanges with the afterlife—or sometimes simply entertained them, as in the case of the spirit Jim whose black hand is materialized in the opening séance scene of William Dean Howells's *The Undiscovered Country* (1880). Unable to "moderate his transports," Jim slaps the sitters, to their delight.[29]

Sally's phantasmal mischief is most evident in the chapter titled "Sally Plays Medium (Subconscious Writing)," which focuses on the time when BIV, gazing in a mirror, abruptly sees her face take on a "curious laughing expression":

> It was not her own expression, but one that she had never seen before. It seemed to her devilish, diabolical, and uncanny, entirely out of keeping with her thoughts. (This expression I recognized from the description to be the peculiar smile of Sally, which I had often seen upon the face of BI or BIV.) IV had a feeling of horror come over her at what she saw. She seemed to recognize it as the expression of the thing that possessed her. . . . It suddenly occurred to her to talk to this "thing," to this "other person," in the mirror; to put questions to "it.". . . Accordingly, placing some paper before her on the bureau and taking a pencil in hand, she addressed herself to the face in the glass. Presently her hand began to write. (361)

"Possessed" by an otherworldly, "thing"-like personality, BIV proceeds to automatic writing, through which Sally at first professes to be a "spirit." Prince details such anecdotes to a degree that clearly exceeds their scientific value, purposefully echoing fictional and (purported) non-fictional stories of haunting. "I am certain," Prince prefaces this incident, ratcheting up its suspensefulness, "that in Biblical, and perhaps in early Salem days, when they hung witches, if a person exhibited the peculiar manifestations which I am going to relate she would have been considered to be literally possessed by the devil" (360). But the cultural touchstone he has in mind is obviously as much modern spiritualism as the Salem witch trials. In another instance we are told that in automatically writing through Miss Beauchamp, Sally "thought it in

---

29. William Dean Howells, *The Undiscovered Country* (1880; Boston: Houghton, Mifflin, 1896), 27.

keeping with her part to imitate the style and manner of the mediums. She found it much more effective, she said, to cross her t's and to dot her i's, and to write in rather a scraggly way as the mediums do" (430). This just follows a spectral description of the hallucinations that, in her subconscious capacity, she causes to plague BI and BIV: "Sally did not rest content with this material world, but let her imagination run riot in the realms of the supernatural. She made them see faces and hands at the window,—not real faces and hands, but ghostly ones. Both BI and BIV were afraid to go to sleep" (429–30).

Prince figures Sally's occult occupation of Miss Beauchamp's body in another way when he writes of his initial detection of Sally's existence. Under hypnosis, Miss Beauchamp has occasionally been referring to her waking self as "she," as if the hypnosis had uncovered a personality different from the primary one. At other times, though, hypnosis has brought only BII, the hypnotic self, who identifies with the waking Miss Beauchamp. Prince asks BII about her periodical lapses, those moments when she has alleged a distinctness from Miss Beauchamp, but because he is really unwittingly talking about Sally, BII does not understand his questions.

> "Listen: now you say you are Miss Beauchamp."
> "Yes."
> "Then why did you say you were not Miss Beauchamp?"
> [Surprised.] "Why, I never said so.". . .
> "Well, you know who you are?"
> "Yes, Miss Beauchamp."
> "Exactly. You have got over the idea of being different from other persons,—that there is a 'She'?"
> [Surprised and puzzled.] "What 'she'? I do not know what you mean."
> "Yes, you do."
> "No, unless you mean Rider Haggard's 'She.'" (28; brackets in original)

The theatrical documentation of BII's confusion (complete with stagy emotional cues) conflates Sally with the ageless African queen of Haggard's popular romance *She* (1887). The effect is not only to underscore Sally's magical character in Prince's narrative, but also to frame his hypnosis as an instrument of discovery along the lines of Holly and Leo's of Ayesha.[30]

---

30. The moment is redolent of what Patrick Brantlinger calls "imperial Gothic": a turn-of-the-century mode of writing in which the occult and atavistic, embodied by an external creature or the

Further, Prince implicitly racializes Sally here: like Ayesha, Sally is an exotic who yet paradoxically wields her powers from the position of a white woman's body. She easily slips into the stock role of foreign degenerate in her irrationality—she has for instance no sense of time: "A day, a week, a month are almost the same to her" (153)—and in her arrested development: "Of course it is manifest that one of the most marked peculiarities of Sally's personality is its *childlike immaturity.* Sally is a child" (152). (Recall another doctor, Van Helsing in *Dracula,* who insists his Oriental foe has only a child's brain.) The occult had by this time been repeatedly linked to non-white and non-Western peoples, in sites ranging from the late Victorian Gothic to spiritualistic discourse; American Indians, for example, made frequent visitations within modern spiritualism.[31] The dark-skinned visitant's difference from the "civilized" world could translate as either spiritual wisdom or, as in the case of the playful Jim in Howells's novel, immaturity. Sally's exoticization through the *She* allusion enhances her position in Prince's narrative as a supernatural presence, a sorceress or an apparition.

To be sure, *Dissociation* at some moments reclassifies apparent magic. Prince indicates that the "'demoniac possessions' of the middle ages" are equivalent to the "hysterical crises of modern medicine" (18) and devotes an entire chapter to the subject of people who confound mental dissociation with the miracles of religious epiphany, in a manner that typifies turn-of-the-century medical approaches to mysticism.[32] But neither his medical perspective nor his real commitment to presenting key data of the case prevents him from stylistic trimmings that suggest Miss Beauchamp's mediation of some wholly discrete, preternatural being. It is possible that these trimmings subtly influenced the interpretation of William McDougall, a psychologist who served important offices in the SPR, that Sally was so distinct in memory, thought, and experience from Miss Beauchamp as to be in reality another soul communicating with the patient telepathically. This kind of soul-to-soul phenomenon, McDougall wrote in the *Proceedings of the Society for Psychical Research,* "would give very strong support to the spiritistic explanation of such cases as [the séance medium] Mrs. Piper, and would go far to justify

---

unconscious itself, fuel masculine exploration and adventure. *Rule of Darkness: British Literature and Imperialism, 1830–1914* (Ithaca: Cornell University Press, 1988), chap. 8.

31. Werner Sollors, "Dr. Benjamin Franklin's Celestial Telegraph, or Indian Blessings to Gas-Lit American Drawing Rooms," *American Quarterly* 35 (1983): 458–80.

32. See Cristina Mazzoni, *Saint Hysteria: Neurosis, Mysticism, and Gender in European Culture* (Ithaca: Cornell University Press, 1996), chap. 1.

the belief in the survival of human personality after the death of the body. It is for this reason that Sally Beauchamp seems to me of so great interest to this Society."[33]

Crafting Personalities

One consequence of Prince's embellishments is that they underscore Sally's difference in temperament from the other personalities, most markedly BI, whom Prince interestingly collapses with his patient *tout court*. "Let it be understood, then," he clarifies, "that in describing Miss Beauchamp, or whenever referring to her, the name, unless otherwise qualified, refers to the first Miss Beauchamp, the one who first came under observation" (7). Because mentions of "Miss Beauchamp" naturally lead us to picture the patient in her entirety, Prince's usage means the woman as a whole seems to acquire BI's traits. In turn, his depictions of Sally's devilishness only magnify her invading, foreign presence in the body of the saintly Miss Beauchamp. On the one hand, BI has a "love of truth" (10) and charitably visits the elderly and is "very fond of church and of church services" (291), "very dependent" (292), and "very fond of children" (293). In short, she embodies true womanly reserve taken to a sickly extreme. On the other hand, Sally is antagonistic, masculine, and fervently anti-domestic. As a young girl, Prince narrates, Sally—whom he initially calls Chris, short for Christine—longed to "play 'hookey,'" thinking it "would be awfully exciting because the boys did it and were always telling about it" (373). As an adult she listens jealously to the sound of men out on the streets: "I envied the men their freedom, and their clothes, and their ability to do as they wished" (340). She also boasts a partiality for French novels and has a French friend with a "distasteful" "foreign point of view" (478), and in general chafes against strict morality—"Do poodles have morals too? Isn't there *any* place free from them?" (356). Most dangerous is her persistence, despite the other selves' disapproval, in carrying on a liaison with the infamous Jones. And she hates babies (387); one of her many pranks involves unraveling a baby blanket Miss Beauchamp has been knitting for a friend (161–62).

The other personality Prince detects also noticeably opposes the kindly BI. BIV is willful, impatient, and irreligious and dislikes the objects of BI's

33. W. McDougall, "The Case of Sally Beauchamp," *Proceedings of the Society for Psychical Research* 19 (1907): 430.

caretaking, the ill and children. If Prince marks Sally's dispositional boundaries through metaphors of possession, he makes the contrast between BI and BIV perhaps even harder to miss by providing a several-page table listing BI and BIV's tastes and "MORAL CHARACTERISTICS" side-by-side, displaying the one's considerate docility and the other's egoism (291–94). This graphic touch reinforces what I would argue is Prince's careful construction of character and, specifically, his exploitation of contemporary gender debates to create narrative conflict and intrigue. Michael Kenny suggests a significance in the predominance of the "woman question" when cases like Miss Beauchamp's were unfolding; Elaine Showalter likewise characterizes the second selves of Prince's patients (Miss Beauchamp as well as another called "B.C.A.") as reactions to confining notions of femininity.[34] It is inviting, in other words, to read BIV and Sally as Miss Beauchamp's acting out of subversive aspects of her personality. But I suspect that Prince himself is doing the inviting, carefully shaping the text, deliberately highlighting the battle over gender identity within his patient by ostentatiously distinguishing the accepted feminine versus anti-feminine strains of her character. The purpose may to some degree be theoretical—related to Prince's notion, expressed in other writing, that women are socially pressured to stifle certain "impulses and cravings."[35] But his representations are so emphatic as to indicate another motive as well, one based in a desire for a broad readership: to produce dramatic conflict by leaning on the controversial and therefore titillating spectacle of modernizing womanhood.

We are never allowed to forget the institution the modern woman jeopardizes—the female-managed family—because Prince remarkably often styles his patient's cluster of selves as just that, a family, the "Beauchamp family." He quotes letters that offer "insight into the condition of the family relations" (425); he tells us about an "outbreak of family rows which compelled the postponement of the study" (309) and about the importance to his treatment of "family concord" (401). A chapter focusing on the personalities' infighting is called "A House Divided against Itself." These metaphors—like the BI/"Miss Beauchamp" fusion, which rhetorically prioritizes BI's heightened feminine virtue—point up a stable domestic identity thrown out of joint, while generating drama out of its elusiveness.

Hints that Prince has inflated the contrast between the personalities are also visible in odd snippets of Sally's letters, which insinuate that privately

---

34. Kenny, *Passion of Ansel Bourne,* 130; Elaine Showalter, *Sexual Anarchy: Gender and Culture at the Fin de Siècle* (New York: Viking, 1990), 121.

35. Quoted in Showalter, *Sexual Anarchy,* 120.

Miss Beauchamp harbors unseemly thoughts and desires. "You are sneaky, just sneaky," she rails to Miss Beauchamp, "and if I were you I'd confess my own sins before I began bewailing those of other people" (126). And again: "you are *proud,* yes, and *honorable,* and quite shocked at your humble servant's disregard of conventions. How I hate you for a hypocrite!" (166). Another letter, this one to Prince, urges him to dig deeper to comprehend Miss Beauchamp's real character: "Just because she's ashamed of certain tastes she has—tastes which are really her salvation—she fancies that 'Sally is cropping out again.' Why don't you read her journal? Why don't you see what's wrong with her, instead of blaming me for everything? Am I the only person who likes to be amused? the only one who finds men interesting? How perfectly silly to blame me for what she does!" (560). Although he includes these letters, Prince pays little attention to their innuendoes, which complicate the dichotomy he implies elsewhere.

Of course one might view Sally's claims as an expression of what Prince reports as her habit in letters to Miss Beauchamp's friends of "exaggerating and distorting [matters] beyond recognition, and even telling things not true" (163). It is also possible, though, that this representation of her letters conceals the exaggerations and distortions of Prince's own writing. He says elsewhere that he relies on Sally, with her subconscious access, to clue him into his patient's nature: "Often it was Miss Beauchamp's own meditations and emotions which were the offending factors [in her nerve strain], and on these Sally kept me well posted. Miss Beauchamp...was exasperatingly reticent about such matters, and kept me in the dark about much that it was important to know. Sally's information, which always proved to be correct, was a great help" (118). Although elsewhere he finds Sally's shenanigans an impediment, his trust in her here lends credence to her claims about Miss Beauchamp's inner moral life, intensifying the question of what Prince's assertions overlook. In part my shelving of the question of the real source of Clara Fowler's alternate selves has to do with the fact that customary answers, attuned to the seemingly telling split between angelic and New Womanish personalities, would not sufficiently recognize the way the text itself composes this binary. The possibility that Prince has created foiling characters must give a feminist critic pause, suggesting as it does an ironic co-opting of an idea of female liberation for the ends of masculine authorial and professional significance.

As a kind of detective narrative, Prince's text gradually moves from attempts to understand Miss Beauchamp's personalities to "The Hunt for the Real Miss Beauchamp," as Part II is called. And just how does he decide BII fits the bill? It is important to Prince that she possesses the thoughts and memories of BI and BIV combined. But then, so does Sally; after some trouble she eventually lays hold of even BIV's interiority. Prince also stresses the necessity for health and stability: "A normal self must be able to adjust itself physiologically to its environment, otherwise all sorts of perverted reactions of the body arise..., along with psychological [symptoms]..., and it becomes a sick self" (233). BI fails this criterion at first glance, while BIV presents more oblique tokens of instability. Here, though, besides Prince's odd snubbing again of Sally (who is never ill), it is hardly evident that BII personifies the kind of resilience Prince wants. In the final chapter, "The Real Miss Beauchamp at Last, and How She Was Found," Prince affirms that this personality,

> however, is not permanent. She has the same emotional psychical make-up which is so prominent a trait in I and IV, and though it is not so intense as in the disintegrated selves, still it is sufficient to be a disturbing factor.... The mental cohesion of a person with such a temperament necessarily yields to the disintegrating effects of the strains of life. (521)

In other words, the Real Miss Beauchamp continues to dissociate. The best Prince can record is a six-month period in which only one "slight lapse" occurred, and even this level of stability was contingent on a careful pruning out of possible strains. In general "she requires a therapeutic suggestion at regular intervals to offset the wear and tear of her life." Prince is troubled enough by the situation to ask, "How far and for how long can she be protected?" (524). But we may well ask differently: has any real cure been achieved?

The volatility of the Real Miss Beauchamp suggests a desperate, ultimately problematic attempt to provide narrative resolution to a tale of discovery. This is not necessarily to say that Prince tailored his basic conclusions in the case to a reading audience. There may have been extratextual motives tending toward his favoring of BII. It is noteworthy that she alone offers no hindrances to Prince's therapy and research. BI is annoyingly reticent and private, while BIV is stubborn and offended by his meddling; she becomes upset when she learns that Prince is writing a book and has spoken about her before the Paris Psychological Congress (453). And Sally is just too mischievous, besides being afraid of her own demise, as well as suspicious that Prince's ideas are erroneous and self-serving. "I do not choose to be psychologized by anyone," she petulantly exclaims (317), and she accuses him of "having a lot of theories that you fit people to, regardless of what those people really are. It's *always* your theories you have in mind—not at all the people" (560). Conversely, BII is remarkably cooperative and indeed "beg[s] pathetically" for

Prince to "help her, protect her from herself, from every one" (434). In addition, not only does she impart freely all she knows about Miss Beauchamp, but she has no qualms at all about Prince's sharing this knowledge with others: "As to the publication of her case, I may use any material that I may have in any way that I please. She does not care at all" (454).

Prince's desire to control the circumstances of his study may play an important part in his act of apparent detection. I would further propose that his hypnosis resembles his narrative-writing in blending the premise of detection with the reality of inventiveness and manipulation. Ruth Leys has drawn attention to Prince's efforts to forestall charges that Miss Beauchamp's personalities amount to artifacts produced by way of hypnotic suggestion and imitation. He wants to insist on their autonomous subjectivity, although, Leys concludes, the text contains evidence that contradicts him. Ironically Prince's claim is weakest in my view as regards BII, because she out of the four personalities depends most for her survival on the hypnotic setting. Not only is her identity predicated on hypnosis; she is the sole personality never to appear on the scene without Prince's trance agency. In the case of the others, he may make them alternate through hypnosis, but each more often takes over the patient's body by herself. Only BII must always be summoned, as Prince himself says: "It is necessary to bring BII by word of command" (302). These facts cast more doubt on the validity of the Real Miss Beauchamp. Though he has framed it as a means of revelation, hypnosis may actually allow him to fashion a new personality. If Prince's psychical imagery and other textual trappings put him in the literary business of constructing character, that construction seems to take on greater empirical resonance in his hypnotic procedure.

The causative, and by extension coercive, aspect of Prince's hypnosis is strikingly implied in the following scene, in which he attempts, against BIV's strong opposition, to hypnotize her into the deepest state possible, the state that brings BII.

It was a battle of wills. She fought mentally and physically. She shrank from me and ran to the farther end of the room, endeavoring with averted face to avoid my eye. She tried to escape the grasp of my hand on her wrist, feeling, as she did, the intensification of the hypnotic influence by the physical touch. . . .

There was over her spine a "hypno-genetic point," pressure upon which always caused a thrill to run through her that weakened her will and induced hypnotic sleep. I pressed upon this point. As she felt the hypnotic influence coming over her, her will-power weakened. With a strength that I did not think she possessed she fought, wrestled, struggled to throw off my grasp, and to resist the subtle power. The rush of physical sensations through her body, and the mental feeling that her consciousness was being engulfed in oblivion were more than she could bear. In mental anguish she shouted aloud. Thus for two hours we struggled. (493–94)

The encounter transmutes hypnotic into sexual intimacy, even ravishment, complete with heightened bodily responses, an overwhelmed consciousness, and ecstatic shouts.[36] Fears of sexual predation—the usually male control's exploitation of his female subject—had hovered around Victorian tales of hypnosis, especially by popular entertainers.[37] Prince's struggle to overmaster BIV suggests that the field of psychology licensed such incursions in the name of science.

The moment recalls a discussion in *The Psychology of Suggestion* (1898), by Prince's colleague Boris Sidis, of the work of French hypnotist Alfred Binet with his hysterical patient Elise B. Having commanded the hypnotized Elise that she will not perceive him upon waking, Binet begins uttering scandalous statements to test if these will turn her attention toward him:

"You can not deceive me. It is only two years since you had a child, and you made away with it. Is that true? I have been told so." She did not move; her face remained peaceful. Wishing to see, on account of its medico-legal bearing, whether a serious offence might be committed under cover of a negative hallucination, I roughly raised her dress and skirt. Although naturally very modest, she allowed this without a blush. I pinched the calf of her leg and thigh. She made absolutely no sign whatever. I am convinced that she might have been assaulted in this state without the slightest resistance.[38]

The hypnotic woman's pliability means she offers multiple gratifications to the medical man who wants to know her inside and out.

---

36. Commenting on Prince's violating maneuvers, Rosenzweig glosses the "hypno-genetic point" as a reference to contemporary hypnotists' theory of "hysterogenic" or "erogenic" zones, pressure upon which was said to stimulate genital arousal, even orgasm ("Sally Beauchamp's Career," 51–52).

37. Daniel Pick, *Svengali's Web: The Alien Enchanter in Modern Culture* (New Haven: Yale University Press, 2000), chap. 5.

38. Quoted in Boris Sidis, *The Psychology of Suggestion: A Research into the Subconscious Nature of Man and Society* (New York: Appleton, 1898), 111.

Affirming the rationalistic ("medico-legal") credibility of hypnotism required overlooking its uncertain relationship to knowledge. The reliability of its disclosures was far from clear, as it could be used as much to command and shape as to probe the unconscious will of another. Prince arguably forms the substance of his discovery of the Real Miss Beauchamp, along with the contest over notions of femininity that seems to underlie his patient's dissociation. If, amid all this, it is difficult for us not to continue to ponder the conscious designs the other real Miss Beauchamp—Clara Fowler—might have had in manifesting multiple personalities, our thoughts echo the plotlines of those detective narratives, Gallon's and Sayers's, that treat as dubious (doubled, two-faced) the picture of women as simply others' inattentive means to know. As a genre, the detective narrative, whether a literal tale of crime-busting or a more figurative instance of occult uncovering, foregrounds the path toward knowledge. Examples featuring women as detective devices point up potential complexities along that path: the obscurity of who owns what is known and communicated and sometimes, concomitantly, the indefinite validity of that knowledge. Whether or not we feel confident about the status of what is revealed may depend much on our belief in women's easy personal and mental abstraction. Interestingly, not even Doyle seems wholeheartedly to believe in this: for all the revisionism of "A Case of Identity," Mary Sutherland does not negate the memory of Irene Adler. Several narratives analyzed in this study indeed put forth at best qualified depictions of female unconsciousness. The automatic woman may well have served, at the core of turn-of-the-century knowledge and communication networks, as an object of faith—like all such objects, a hoped-for reality that will not bear significant scrutiny.

CHAPTER 5

Literary Transmission
and Male Mediation

> Curse your women too,
> Your insolent wives and daughters, that fire up
> Or faint away if a male hand squeeze theirs,
> Yet, to encourage Sludge, may play with Sludge
> As only a "medium," only the kind of thing
> They must humour, fondle...oh, to misconceive
> Were too preposterous!
>
> —Robert Browning, "Mr. Sludge, 'The Medium'"

Sarcasm and humiliation simmer through Mr. Sludge's confession of phony spirit mediumship in Robert Browning's 1864 dramatic monologue. Especially infuriating to Sludge is the un-gendering effect of his work: because he is always pretending to be someone else, his grasp would not be perceived by séance sitters as a sexual threat, and women merely infantilize him or, at best, use him for an erotic thrill they refuse to give him credit for. Rhetorically slyly, but probably accurately, Sludge blames his addressee Hiram H. Horsefall and Horsefall's genteel friends for his own deceptions: it is they who have turned him into a "showman's ape," egging on every antic and

> any mood
> So long as the ape be in it and no man—
> Because a nut pays every mood alike.[1]

If Sludge feels any gratitude toward them for their patronage, it is only

> The gratitude, forsooth, of a prostitute
> To the greenhorn and the bully—friends of hers,

---

1. Robert Browning, "Mr. Sludge, 'The Medium,'" in *The Poems*, ed. John Pettigrew, vol. 1, 821–60 (New York: Penguin Books, 1993), ll. 600, 602–4. All further line number references to "Mr. Sludge, 'The Medium'" are parenthetical within the text.

From the wag that wants the queer jokes for his club,
To the snuff-box-decorator, honest man,
Who just was at his wits' end where to find
So genial a Pasiphae! All and each
Pay, compliment, protect from the police:
And how she hates them for their pains, like me! (ll. 783–90)

Sludge's styling of himself as an exploited woman suggests the air of questionable masculinity that attached to Victorian men—for example, the much bruited medium Daniel Dunglas Home, on whom Browning's speaker is based—in the position of occult sensitive.[2] For Sludge, even the feint of séance channeling has this effect in that, like authentic mediumistic unconsciousness, if in a different way, it requires a suppression or cloaking of self that prevents others from acknowledging him as a man. From Browning's day on, many texts associate occult receptivity with a deficiency of male strengths; more specifically, it often codes (and sometimes causes) sexual passivity or socioeconomic weakness. We see both types of failing in Sludge: besides the fact that the ladies consider his touch innocuous, denying him sexual agency, there is the suggestion of his lower class (he compares himself to "a poor lad,...a help's son" [l. 98]) and that this is what motivates him to play the ape to the nuts who will pay. The "Mr." of Browning's title is a facetious, double-edged jab at his lack of virility and respectability alike.

The sexual objectification and diminished social power of the male sensitive carries over to late Victorian tales of foreign psychical manipulation. In Arthur Conan Doyle's The Parasite (1894), Professor Austin Gilroy's impressionable, visionary nature predisposes him to entrancement by a West Indian woman who wrenches from him displays of ardor; when he tries to rebel, she retaliates with professional sabotage, compelling him to speak inanities in his class lectures, for which he is ultimately suspended from his post. The first victim of the mesmerizing Egyptian insect-priestess in Richard Marsh's The Beetle (1897) is an unemployed clerk who, after hours of fruitless searching for work and being refused a night's shelter by the local asylum, breaks into a seemingly empty house, where he finds himself the helpless object of the Beetle's repeated physical and moral indignities. "For the time I was

---

2. For Daniel Karlin, Sludge's gender uncertainty reflects Browning's view of Home as an arriviste lacking in manliness; Browning's Hatreds (Oxford: Oxford University Press, 1993), chap. 3. Vieda Skultans notes the stigma of effeminacy attached to male spirit mediums in "Mediums, Controls, and Eminent Men," in Women's Religious Experience: Cross-Cultural Perspectives, ed. Pat Holden (London: Croom Helm, 1983), 23.

no longer a man," Robert Holt recalls, describing the Beetle's automatizing power; "I was, in the extremest sense, an example of passive obedience."[3] Forced to undress, probed and kissed, he is finally dispatched—like Doyle's Gilroy—on a thief's errand that seems to represent the fulfillment of his downward social trajectory.

In Rudyard Kipling's "Wireless" (1902), male spirit possession is freighted with implications of gender upheaval and the sapping of male control. The tale reflects on modern conditions in depicting men's confused subjection to two interrelated, magically liberated forces: women and electricity. These period-conscious themes are the center of this chapter's discussion of "Wireless." However, it is worth noting that what the male medium channels in this story is poetry. This represents a striking near-literalization of the trope in which some form of mediation in the plot—typing in *Dracula,* spiritual journeying in *A Romance of Two Worlds,* mesmerized singing in *Trilby*—serves as an analogue for the mediation of literature itself. "Wireless" makes a show of the channeled communication of John Keats's verses framing literature as a complex, multi-station transmission.

As another work whose in-text mediation figures the narrator/speaker delivering the text from author to reader, take "Mr. Sludge, the 'Medium.'" There is a moment when Sludge mocks his sitters for buying into his ridiculous impersonations of the other world:

I've made a spirit squeak
In sham voice for a minute, then outbroke
Bold in my own, defying the imbeciles—
Have copied some ghost's pothooks, half a page,
Then ended with my own scrawl undisguised.
"All right! The ghost was merely using Sludge,
Suiting itself from his imperfect stock!" (ll. 592–98)

Sludge's parody of his sitters' rationalization of his audacious self-revelation reminds us of the root of his anger, their complete unwillingness to grant him self-presence. But in addition, the moment—specifically its description of the scene of writing—echoes at two later points in his monologue. The

---

3. Richard Marsh, *The Beetle* (1897; Peterborough, Ont.: Broadview, 2004), 54. On the Beetle's aberrantly forceful femininity and emasculated victims, see Kelly Hurley, *The Gothic Body: Sexuality, Materialism, and Degeneration at the* Fin de Siècle (Cambridge: Cambridge University Press, 1996), 124–47 passim. Hurley points to other late Victorian tales indicating a similar relationship between mesmeric females and gender disruptions (187 n. 5).

first point comes when Sludge is trying to convince a doubting Horsefall that
he, Sludge, benefits from providential guidance:

> Oh, you wince at this?
> You'd fain distinguish between gift and gift,
> Washington's oracle and Sludge's itch
> O' the elbow when at whist he ought to trump?
> With Sludge it's too absurd? *Fine, draw the line*
> *Somewhere, but, sir, your somewhere is not mine!*
>
> Bless us, I'm turning poet! (ll. 1178–84; emphasis in original)

The poem's only other instance of italicization, aside from a few words and
phrases, occurs at the second point I have in mind, when Sludge is again trying to dodge Horsefall's charge of deceit:

> What, sir? You won't shake hands? "Because I cheat!"
> "You've found me out in cheating!" That's enough
> To make an apostle swear! Why, when I cheat,
> *Mean to cheat, do cheat, and am caught in the act,*
> *Are you, or, rather, am I sure o' the fact?*
> (There's verse again, but I'm inspired somehow.)
> (ll. 1280–85; emphasis in original)

Arrestingly, both cases of italicization signal end-rhymes, which are anoma-
lous in this work but of course a traditional hallmark of verse, and both are
followed by Sludge's declaration of poetic inspiration.

Given their stylistic abruptness and typographical showiness, which make
them jar with the rest of the monologue, we can read these expressions as
Browning himself peeping out through his work, calling attention to his
poetic authorship, writing in his "own scrawl undisguised." In one sense
the three passages read together in this way invite us to compare Browning
to Sludge, reinforcing the argument that our most celebrated dramatic mo-
nologist, preoccupied with reviving dead speakers, imitates the spiritualistic
practices he ostensibly derided.[4] Yet in another sense, one that concentrates
finally on the second and third passages, Browning is not comparing himself

---

4. Adam Roberts, "Browning, the Dramatic Monologue and the Resuscitation of the Dead,"
in *The Victorian Supernatural,* ed. Nicola Bown, Carolyn Burdett, and Pamela Thurschwell, 109–27
(Cambridge: Cambridge University Press, 2004).

to but rather, just the opposite, distinguishing himself from Sludge, in such a way as to make us aware of the poem as a feat of transmission. By displaying himself as poet in the rhymed lines, Browning vividly if fleetingly effects a separation from Sludge. In so doing he accentuates his work as a layering of selves or, in more linear terms, as a relay from himself to a speaker—which in turn posits the dialogue between Browning as author and us as readers as dependent on intermediation: that, precisely, of Sludge, the medium.

It is the concept of the speaking textual voice as a medium, and particularly of its potential for Sludge-like duplicity, that guides my reading of the psychic sensitive Latimer in George Eliot's *The Lifted Veil* (1859). Here we return to the mid-nineteenth century, when modern spiritualism was relatively new. Browning's monologue, also from this period, gives a sense of how quickly the medium's instrumentality and self-suppression came to be associated with the feminine, and how damning this could be for male practitioners. *The Lifted Veil* implies a similarly rapid gendering of the occult instrument's sensitivity, as this is linked to the protagonist's unfitness for masculine gainful endeavors. Moreover, Eliot's tale suggests a first tentative exploration, soon after spiritualism came to England, of what it might mean to depend on others to mediate our messages. My focus is on how Latimer's narration shapes the story's transmission and how *The Lifted Veil* thus meditates on channeled communications. Both the emasculation commonly linked to occult mediation and the notion of literature as a mediated transmission inform my closing discussion of T. S. Eliot's "Tradition and the Individual Talent" (1919). This author effectively re-genders mediation in his vision of the implicitly male poet-medium. Yet it is a testament to the staying power of mediation's femininity that despite these transformations, the concept of psychic womanhood endures elsewhere in twentieth-century writing and culture, and in fact re-emerges in Eliot's own body of work.

Suspect Communications in *The Lifted Veil*

George Eliot's *The Lifted Veil* is a tale of strange perceptiveness thrust upon a man who only shares the extraordinary secret of his ability with us, his readers. "I have never fully unbosomed myself to any human being," says Latimer; "I have never been encouraged to trust much in the sympathy of my fellow-men."[5] The faculty of "prevision" dawns on him during his convalescence

---

5. George Eliot, *The Lifted Veil*, in *The Book of Fantastic Literature by Women, 1800–World War II*, ed. and intro. A. Susan Williams (New York: Caroll and Graf, 1992), 58. All further page references to *The Lifted Veil* are parenthetical within the text.

after a serious illness. In a state of "rapt passivity," he sees an image of Prague (65); the city is unknown to him, yet the details of his vision are verified when he visits the place in later weeks. Soon he is also entering others' minds—experiencing what Frederic Myers would in a few decades dub telepathy—and feeling tortured by the bland pettiness and conceit beneath the sparkling demeanor of his father, his brother, and their well-mannered friends. Latimer's paranormal susceptibility is presumably the outgrowth of his "morbidly sensitive nature" and "fragile, nervous, ineffectual self" (69, 68). His "half-womanish, half-ghostly beauty" (68) denotes the effeminacy that Victorians often assigned to the nervously debilitated man. These traits have a specific index in terms of Latimer's potential for social productiveness, or lack thereof: his father, a banker, disparages him as a poet-type and tries in vain to give him a practical education.[6] His brother, Alfred, becomes engaged to the blonde, seductive Bertha, who fascinates Latimer as the one individual mysteriously out of reach of his telepathy; his infatuation continues even after he has had a prevision of himself caught with her in a marriage of mutual scorn. After Alfred's sudden death, Bertha indeed becomes Latimer's wife, but when the veil concealing her inner self is at last lifted, Latimer perceives all of her spiritual sterility and vain malice, and their relationship spools out into agonizing years in which each avoids the other's hateful presence.

Latimer's magical access to others' minds has often seemed like a mimicry of authorial omniscience, and by some lights this is Eliot's means for interrogating her own artistic practice.[7] The story does metaphorically investigate the workings of narrative; but I want to delineate that investigation not by interpreting Latimer as a figure for the author, but rather by focusing on what he unambiguously is, a narrator, and by scrutinizing his capacities, goals, and effects in that position. Reading this story, I propose, compels us to think about narration as a mediated transmission. And in that aspect *The Lifted Veil*

---

6. On Latimer's feminine nervousness and its social implications, see also Jane Wood, *Passion and Pathology in Victorian Fiction* (Oxford: Oxford University Press, 2001), 78–109.

7. The premise of Latimer's author-like omniscience more or less informs the following: Gillian Beer, "Myth and the Single Consciousness: *Middlemarch* and *The Lifted Veil*," in *This Particular Web: Essays on* Middlemarch, ed. Ian Adam, 91–115 (Toronto: University of Toronto Press, 1975); Charles Swann, "Déjà Vu: Déjà Lu: 'The Lifted Veil' as an Experiment in Art," *Literature and History* 5 (1979): 40–57, 86; Terry Eagleton, "Power and Knowledge in 'The Lifted Veil,'" *Literature and History* 9 (1983): 52–61; Peter K. Garrett, *Gothic Reflections: Narrative Force in Nineteenth-Century Fiction* (Ithaca: Cornell University Press, 2003), 175–84; and Richard Menke, *Telegraphic Realism: Victorian Fiction and Other Information Systems* (Stanford: Stanford University Press, 2008), chap 4. Neil Hertz argues for the compulsory, authorial nature of Latimer's telling; *George Eliot's Pulse* (Stanford: Stanford University Press, 2003), chap. 4. Julian Wolfreys also notes the critical tendency to see the story as a self-referential examination of narrative; *Victorian Hauntings: Spectrality, Gothic, the Uncanny and Literature* (Houndmills, UK: Palgrave, 2002), 78.

typifies its decade, a period in which mediation was becoming especially visible as a component of communication, occult communication especially.

It is likely that Eliot's interest in mesmerism and clairvoyance informed her depiction of Latimer's preternatural abilities.[8] But perhaps even more sensational than these trends during the 1850s, when she wrote *The Lifted Veil*, was the spread of modern spiritualism to England. The American mediums Mrs. Hayden and Mrs. Roberts created a stir when they came over in 1852 and 1853, to be followed soon after by the arrival of the celebrated Home in 1855.[9] Several letters by Eliot demonstrate her alertness to spiritualism in the years shortly before as well as after the writing of her story.[10] Prevision and telepathy are evident forms of occultism in Latimer's memoir, but certain facets of his experiences call to mind the spirit medium. His glimpse of Prague is of a netherworld made gloomy by the "summer sunshine of a long-past century arrested in its course" and haunted by "people doomed to live on in the stale repetition of memories"; the place teems with "ephemeral visitants" (63). Then, Latimer describes the picture of oversensitivity he knows he presents to Bertha after their marriage as that of a "miserable ghost-seer, surrounded by phantoms in the noonday" (85). Even the manner in which he acquires his abilities—around adolescence and following a severe sickness—evokes the common life history of the Victorian spirit medium.[11] Latimer never acts spiritualistically within the plot, yet in drawing on the theme of occult sensitivity in general and connoting spirit mediumship in particular, Eliot suggests his potential for communication mediation—a potential then realized not within but rather through or across the bounds of the story.

The Lifted Veil first draws our attention to Latimer as narrative medium through his frequent references and direct addresses to the reader: these give the story an emphatic third dimension or axis along which it reaches out to us. We become aware of the narrative path of information, of its movement toward us as its receivers. At the same time, instances of mediation within the tale are metaphors that invite us to consider how Latimer

8. B. M. Gray, "Pseudoscience and George Eliot's 'The Lifted Veil,'" *Nineteenth-Century Fiction* 36 (1982): 407–23. On mesmerism in the story, see also Malcolm Bull, "Mastery and Slavery in 'The Lifted Veil,'" *Essays in Criticism* 48 (1998): 244–61.

9. Janet Oppenheim, *The Other World: Spiritualism and Psychical Research in England, 1850–1914* (Cambridge: Cambridge University Press, 1985), 11.

10. Vanessa D. Dickerson helpfully excerpts these letters within her own analysis of *The Lifted Veil* in *Victorian Ghosts in the Noontide: Women Writers and the Supernatural* (Columbia: University of Missouri Press, 1996), 90–93.

11. Alex Owen discusses pubescent illness and mediumship in *The Darkened Room: Women, Power, and Spiritualism in Late Victorian England* (Philadelphia: University of Pennsylvania Press, 1990), 206–9.

operates along this axis, how he intervenes in our reception of this product of Eliot's imagination. Understanding that premise helps to explain the significance of the famously outlandish transfusion scene near the tale's conclusion. Latimer is visited by his only close friend, the scientist Charles Meunier, who performs an experiment on Bertha's maid, Mrs. Archer, immediately after the latter's death. Meunier's blood transfusion restores Mrs. Archer to life for a few moments, just enough for her to exclaim about her mistress: "You mean to poison your husband...the poison is in the black cabinet...I got it for you...you laughed at me, and told lies about me behind my back, to make me disgusting...because you were jealous...are you sorry...now?" (95; ellipses in original). In this scene the flow of blood is roughly tantamount to the flow of information, for it is the influx of blood that suddenly instigates Mrs. Archer's revelation of Bertha's hidden secret. This symbolic equivalence between blood and information by extension formulates information itself as transmissible—as movable from one site to another and as subject to flows and channels.[12] Simultaneously, it is tempting to see Mrs. Archer here as Latimer's double. Her sickly, feminine body recalls his; and just as she becomes a medium of Bertha's secret, scrambling out of hatred to disclose it before she dies (again), so does Latimer, motivated by his own alienation from Bertha above all others, make a last gasp of an effort to tell us his tale before what he foresees as his fast approaching death.[13] The scene as a whole becomes an allegory for Latimer's own status as a conduit of information— where the story he hastens to tell us ultimately originates with Eliot.

This is not to say that Latimer's relaying of Eliot's narrative is unproblematic, however. While enlarging on occultism as a thematic touchstone in the story, I would argue that we have much reason to question the literal extent of what Eliot's narrator claims to be his abilities. Criticism on *The Lifted Veil* has given surprisingly little sustained attention to the possibility of this dubiousness—to the notion, for example, that Latimer's telepathic knowledge comes down to what Terry Eagleton labels "paranoid intuitions," being a

---

12. The Victorians tended to see information in terms of flows, as Richard Menke describes (*Telegraphic Realism*, 20–21). In his own reading of *The Lifted Veil*, Menke compares the disclosure afforded by Meunier's experiment with the vivisectional, prosaically informative quality of Latimer's visions (*Telegraphic Realism*, 151–52).

13. Mrs. Archer's indictment itself can be seen as a substitute for Latimer's, as Helen Small notes; introduction to *The Lifted Veil/Brother Jacob*, by George Eliot (Oxford: Oxford University Press, 1999), xxiv n. 37. What most intrigues me are the strikingly parallel circumstances of the indictment. Others see a doubling between Mrs. Archer's transfused blood and Latimer's experience of preternatural knowledge; see Kate Flint, "Blood, Bodies, and *The Lifted Veil*," *Nineteenth-Century Literature* 51 (1997): 467–68 and Sally Shuttleworth, introduction to *The Lifted Veil and Brother Jacob*, by George Eliot (New York: Penguin Books, 2001), xxx.

"mere projection of his own arrogance, anxiety and aggressivity" onto other minds.[14] The case for Latimer's unreliability is worth elaborating in detail.

To begin with, here he is telling us about the time he let slip a hint of his telepathy to others:

> [Alfred] had occasionally a slightly-affected hesitation in his speech, and when he paused an instant after the second word, my impatience and jealousy impelled me to continue the speech for him, as if it were something we had both learned by rote. He coloured and looked astonished, as well as annoyed; and the words had no sooner escaped my lips than I felt a shock of alarm lest such an anticipation of words—very far from being words of course, easy to divine—should have betrayed me as an exceptional being... But I magnified, as usual, the impression any word or deed of mind could produce on others; for no one gave any sign of having noticed my interruption as more than a rudeness, to be forgiven me on the score of my feeble nervous condition. (72)

This, an opportunity for other characters to confirm for us Latimer's special powers, gestures instead in the other direction, toward his own invention. For Alfred's "astonished" blush and "annoyed" look probably simply register his shock at Latimer's impoliteness; certainly, as Latimer recognizes, this is the only way in which their companions perceive his interjection. Then there is the moment when the veil is finally lifted for him, as he says, on Bertha's thoughts. Notable here are the circumstances surrounding this event: Latimer has just come from his father's death-bed, and "Perhaps it was the first day since the beginning of my passion for [Bertha], in which that passion was completely neutralised by the presence of an absorbing feeling of another kind" (84). His own description of this moment encourages us to read his sudden understanding of the shallow woman Bertha really is not as supernatural knowledge but rather as the more mundane kind of unveiling that ensues after the first phase of romantic love (and lust) has been tempered by life's more grave distractions.

---

14. Eagleton, "Power and Knowledge," 59, 57. Jane Wood offers a Victorian diagnosis for Latimer's delusions: the nervous disease hypochondriasis, involving unruly imaginings; see also Diane Mason, "Latimer's Complaint: Masturbation and Monomania in George Eliot's *The Lifted Veil*," *Women's Writing* 5 (1998): 398–99. While comparing Latimer to the Spasmodic poets, Kevin Ashby notes contemporary suspicions about mesmerism and clairvoyance and particularly questions Latimer's previsions; "The Centre and the Margins in 'The Lifted Veil' and Blackwood's *Edinburgh Magazine*," *George Eliot-George Henry Lewes Studies* 24–25 (1993): 132–46. With the exception of Ashby's work, considerations of narrative fallibility in Eliot's story have tended to be in passing because heavily subordinated to other issues.

Latimer reports that he partially loses his telepathic ability for a while. If we presume he is misinforming us, we can read this development as a sign of his emotional exhaustion: married for seven years by this point, he has become so jaded about his wife's character as to be indifferent to, instead of apprehensive about, what she might be thinking. Now he bears even oddities in her behavior "with languid submission, and without feeling enough interest in her motives to be roused into keen observation" (89). (Significantly, though, his telepathy/apprehensiveness will return at the end of the story, after he has discovered Bertha's murder plot, as well as been compelled by their separation into an uncomfortable "dependence on his servants" [95].)

But according to Latimer, Bertha has picked up on his mind-reading, so that after he utters something that makes obvious his loss of it, she notes with pleasure, "I used to think you were a clairvoyant, and that was the reason why you were so bitter against other clairvoyants, wanting to keep your monopoly; but I see now you have become rather duller than the rest of the world" (90). At first Bertha's statement seems to corroborate his abilities, but her allusion to Latimer's animosity "against other clairvoyants" is puzzling, because no one else claiming occult abilities appears in the story. Hence it is less likely that Bertha is acknowledging a mind-reading capacity in Latimer than that she is using "clairvoyant" somewhat loosely, to encompass anyone with the ability to "see" or understand people or situations with clarity. And in that case she is probably counting herself foremost among the other "clairvoyants" whom Latimer resents for their insight into his own pitiful nature. From their first acquaintance, Bertha and Latimer's relationship has taken its odd tone from a reciprocal "clairvoyance": each has always known the other's faults—he, her cold disparagement, and she, his unsuitability for any life outside a poet's dream. Bertha implies that it is this reciprocity—Latimer's want of a "monopoly"—that disturbs him, because it confronts him with her judgment of him.[15] Now, however, she is delighted to find him uncurious about her, no doubt because this serves rather well her recent homicidal bent.

Once we have doubted Latimer's telepathy, his previsions begin to seem suspicious, too. Latimer himself inadvertently offers the basic principle that allows us to explain these events more logically. Describing how his desire for Bertha persists even after his horrible prevision of her, Latimer begs the reader to try to understand his painful state of inner chaos: "Yet you must

---

15. Eagleton may be thinking of the passage I discuss here when he writes of an earlier scene in which Latimer describes his and Bertha's simultaneous judgment, "The worst blow is that he lacks a monopoly on 'clairvoyance'" ("Power and Knowledge," 58).

have known something of the presentiments that spring from an insight at war with passion; and my visions were only like presentiments intensified to horror" (75). Defining Latimer's prophecies as "intensified presentiments" does put them within the range of our understanding, so much so in fact that they lose much of their mystery. The notion that Latimer's narrative simply distorts to supernormal proportions common human tendencies especially rings true in the case of his prevision of himself and Bertha as a married couple. Given what he admits to be his recklessly irrational desire—even before his prevision he seriously considers his handsome and sociable brother as his rival, refusing to believe Bertha will ever marry Alfred and holding out hope that she will choose him instead (69–71)—it is hardly surprising that he would have a vision of her as his wife. Neither is it surprising that such a vision would be tinged with fear in view of his frailty as measured against what he quite early on perceives to be her haughtiness, and the profound incompatibility of their interests. Even while proclaiming to be at first in a thrilling "state of uncertainty" about her thoughts, soon after meeting her he is already summing her up as "keen, sarcastic, unimaginative, prematurely cynical," and "especially contemptuous towards the German lyrics which were my pet literature at that time" (69).

Importantly, moreover, Latimer's supposed prevision is immediately preceded by his unsettling viewing of a painting of the "cruel-eyed," "cunning, relentless" Lucrezia Borgia. The work has the effect of vaguely impressing upon him the utter self-destructiveness of his passion: "I felt a strange poisoned sensation, as if I had long been inhaling a fatal odour, and was just beginning to be conscious of its effects." Bertha's approach in the garden of the museum just afterward brings to a head the redolent associations of the painting; as she takes his arm, "a strange intoxicating numbness passed over me, like the continuance or climax of the sensation I was still feeling from the gaze of Lucrezia Borgia." Within the vision that follows, Bertha's insulting epithets and "bitter smile of contempt" simply play out his worst dread (73).

Bertha does not speaks these epithets aloud, either in the vision or when the event it foretells occurs, making it easier to write them off as another instance of his dubious narration. Yet it seems harder to explain away Latimer's previsions of certain concrete details—the serpent brooch Bertha will wear in this scene, the patch of rainbow light in Prague, the date of his own death. Regarding the latter, though, Jane Wood tells us that Victorian skeptics dismissed alleged clairvoyant prophecies of death like Latimer's as matters of self-suggestion in nervously disordered minds.[16] Latimer has already learned

---

16. Wood, *Passion and Pathology,* 101.

from his doctor that he suffers from *angina pectoris* and cannot live much longer; that he would use that information to will his own death squares with his wish to be rid of "the wearisome burthen of this earthly existence" (57). Further, as Wood proposes, Latimer's nervous symptoms point to a Victorian medical condition associated with excessive imagination, and this condition, coupled with his haste to remember his life story before dying, compromises his status as a narrator.[17] A diagnosis of pathological inventiveness would help to explain the vivid details of Bertha's brooch and the rainbow light in Prague—especially once we stop to consider that while he is calling this kind of thing a prevision, he is actually narrating it after the fact.

Latimer may indeed have a perilous memory. Just before telling of the transfusion, and apparently trying to account for his lack of insight into Bertha's murder plot, he asserts the diminishment of his powers, plus diminished contact: "My interviews with my wife had become so brief and so rarely solitary, that I had no opportunity of perceiving these images [of her machinations] in her mind with more definiteness." But then he goes on: "The recollections of the past become contracted in the rapidity of thought till they sometimes bear hardly a more distinct resemblance to the external reality than the forms of an oriental alphabet to the objects that suggested them" (88). Latimer's abrupt shift in verb tense would seem to indicate that he is referring to his current scene of writing, attributing the fault in his narrative not just to the past but to the present confusion of his recall. That admission has the potential to alter our whole sense of his memoir. At the very least it prompts us all the more to wonder if his most amazing premonitions are explicable as misrememberings shaped retrospectively by an awareness of later events. Even his first vision of meeting Bertha before they actually meet may be rationalized in this way, as a fanciful mnemonic "contraction" of an "externally real" occurrence with its presentiment.

Ethical Distortions of the Unreliable Medium

One advantage of interpreting Latimer's paranormal faculties as false is that doing so returns *The Lifted Veil* to a realist mode, resolving the story's seeming eccentricity within George Eliot's canon. If, nevertheless, his flawed mental processes have proven easy to overlook, his moral deficiencies have been much less so. Latimer exhibits the type of unreliability James Phelan has termed *misregarding*, in which the narrator's ethics strike us as out of sync with the

---

17. Ibid., 98.

text's implied value system.[18] In Latimer's case, the fault turns on the exercise of sympathy. Eliot is demonstrating her well-known preoccupation with this state of feeling in its spurious formulation within Latimer's narration. He ironically reverses the emotion: it fuels his egotism rather than his altruism; he sees himself only as sympathy's recipient, hardly ever as its giver. His narrative is punctuated with more or less explicit cries out to the reader in this vein, as in his petulant opening statement, "It is only the story of my life that will perhaps win a little more sympathy from strangers when I am dead, than I ever believed it would obtain from my friends while I was living" (58–59). Later he interrogates us with, "Are you unable to give me your sympathy—you who read this?" (75).[19] In this light, his "clairvoyance" begins to look like a mere parody of sympathy, involving a perceptiveness about others that, however, disparages instead of takes to heart their feelings and interests.[20] Notably, seeing into his brother does not further but rather precludes a sense of sympathetic community. His "exasperating insight into Alfred's self-complacent soul," by its contrast with his own unbearably hypersensitive nature, seems from Latimer's perspective "to absolve me from all bonds towards him" (78).

The narrator's sympathy rests on a foundation of antagonism that pollutes all his relations in the story, and that he then carries over to his relation with us, his readers, such that we can only sympathize with him by feeling antipathy toward others. Take as a comparison his befriending of Charles Meunier during their youth in Geneva. Latimer maintains he was attracted by a "community of feeling" between them—but then clarifies that feeling: "Charles was poor and ugly, derided by Genevese *gamins,* and not acceptable in drawing-rooms. I saw that he was isolated, as I was, though from a different case, and, stimulated by a sympathetic resentment, I made timid advances towards him" (62). Here is the same limited, churlish kind of sympathy he seeks with his reader: so much of his life story focuses on people who have neglected or hurt him that one suspects his main goal in telling it is to cultivate a shared animosity. Eventually he appears to lose faith even in the reader,

---

18. James Phelan, *Living to Tell about It: A Rhetoric and Ethics of Character Narration* (Ithaca: Cornell University Press, 2005), 51.

19. On Latimer's opportunistic view of the reader as well as other characters as agents of sympathy for himself, see also Ellen Argyros, *"Without Any Check of Proud Reserve": Sympathy and Its Limits in George Eliot's Novels* (New York: Peter Lang, 1999), chap. 3. Contrast the paradigm of tolerant, helpful sympathy we usually have in Eliot's fiction, detailed in Elizabeth Ermarth, "George Eliot's Conception of Sympathy," *Nineteenth-Century Fiction* 40 (1985): 23–42.

20. For other views of the relationship between sympathy and occult ability in Eliot's writing, including *Daniel Deronda,* see Dickerson, *Victorian Ghosts,* 93–95; Nicholas Royle, *Telepathy and Literature: Essays on the Reading Mind* (Oxford: Blackwell, 1990), chap. 5; and Thomas Albrecht, "Sympathy and Telepathy: The Problem of Ethics in George Eliot's *The Lifted Veil,*" *ELH* 73 (2006): 437–63.

moving from an attitude of presumptuousness—"I shall hurry through the rest of my story...When people are well known to each other, they talk rather of what befalls them externally, leaving their feelings and sentiments to be inferred" (83)—to a suggestion that the reader just may not get it. "We learn *words* by rote, but not their meaning; *that* must be paid for with our life-blood, and printed in the subtle fibres of our nerves," he comments, as if to imply that we cannot possibly commiserate with his suffering merely by reading about it. His next sentences are peevish and vaguely reproachful: "But I will hasten to finish my story. Brevity is justified at once to those who readily understand, and to those who will never understand" (87).

Intriguingly, spirit mediums also in their own way "misregarded" sympathy—at least they would have done so in Eliot's view—and this is another sense in which Latimer evokes contemporary séance feats. Conversations between mortal and spirit worlds were said to depend on the medium's nervous and psychical sensitivity to the will of the spirits seeking to communicate through her, a faculty sometimes spoken of as sympathy.[21] But as we might expect given Eliot's rationalistic tendencies, she herself never believed in spiritualism as an actual phenomenon. Indeed, she seems to have been quite dismayed by what she saw as the deception of it all. Writing in the 1850s, she deplored people she had heard had been gulled into belief; she called spiritualism an "odious trickery" in 1860 and the "lowest charlatanerie" in the decades that followed.[22] When she and George Henry Lewes attended a séance in 1874, they ended up leaving repulsed because the medium would only work in an entirely unlit environment.[23] Eliot singled out Daniel Dunglas Home for special censure: "I would not choose to enter a room where he held a séance. He is an object of moral disgust to me."[24] I am interested in how what Eliot perceived to be the bad faith of someone like Home resembles or models Latimer's. For there is a parallel between the moral disgust she felt for Home and that we are made to feel for Latimer, in that both are responses to a profanation of the concept of sympathy and to the manipulations in the field of communication resulting from this profanation. In calling out to the reader for a sympathy he himself refuses to exercise, Latimer, like the fraudulent spirit medium, implies his comprehension of other-directed feelingness as a basis for community and communication, while in reality twisting it in the service of his own interests.

---

21. See, e.g., Owen's discussion of the "sympathetic vibration" reportedly at work in one family's séance (*Darkened Room*, 83).

22. Quoted in Dickerson, *Victorian Ghosts*, 91.

23. Rosemary Ashton, *George Eliot: A Life* (London: Allan Lane, 1996), 335.

24. Quoted in Dickerson, *Victorian Ghosts*, 91.

I am also intrigued by the way not only Latimer's "misregarding" of sympathy but also his other modes of narrative unreliability highlight the complexities of mediated communication itself—of what happens when one figure is given the task of relaying someone else's message. Consider the fact that even as Latimer longs explicitly and repeatedly for sympathy, Eliot depicts him as a remarkably *un*sympathetic character. Consequently, what Latimer says and what Eliot wants us to understand are at cross purposes: author and reader come to a mutual knowledge through but also, oddly, *in spite of* the narrator-medium's intervention. When we dwell, as narrative theorist Tamar Yacobi does, on narrative as a communicative transmission and the narrator as the medium of that transmission, it is easier to observe the "paradoxical position" of every narrator: he or she acts as a "bridge" between author and reader, but also as an interruptive "wedge" between them; and this paradox intensifies when the narrator is unreliable.[25]

Looking at this situation as applied to Latimer, I would again draw some comparisons between him as a narrator and the fraudulent spirit medium. Both of these figures communicate in a way that, so to speak, lies about what it communicates. This duplicity may be either inadvertent or deliberate, but the effects are the same: a message that means something other than what it says it means. In addition, in both cases, untrustworthiness makes us unusually alert to how the medium's own personality and agenda condition what he communicates. Just as Home's presumed communicative hoax provokes Eliot to evaluate his moral character, so too does our realizing that Latimer is not an uncomplicated conduit of another's thoughts go hand in hand with our sizing him up as a willful and knowing agent in his own right. Of course, we are always attuned to first-person narrators as to varying degrees individuated personalities, distinct from their authors. But when the narrator is unreliable, personality becomes even more solid and conspicuous—not within the plot, but within the channel of communication itself, because it has become an obstacle to our understanding that communication.

Duplicity or untrustworthiness, in other words, thickens the medium's presence within the act of communication. And in literature, it makes us conscious of the medium *as* a medium—as a figure interposed between us and the author, a constant we are normally prone to forget.[26] *The Lifted Veil* is unusual for George Eliot in being a first-person character narration; I would argue that this aberration clues us in to a purposeful formal exploration on her part. Latimer's

---

25. Tamar Yacobi, "Narrative Structure and Fictional Mediation," *Poetics Today* 8 (1987): 335.

26. Yacobi puts the effect in these terms: whereas reliability "is nothing but a contextual neutralization of the mediation-gap" between author and reader, unreliability "is a contextual realization of the mediation-gap" ("Narrative Structure," 336).

thickened, obstructive presence in *The Lifted Veil* connotes Eliot's vigilance toward the mediating position. Vigilance was necessary in the real world, and for many others besides Eliot, because the medium might well challenge the presumed nature of the communication. The peculiar paradox of Latimer's narration shadows the growing awareness of, even trepidation about, people who assume the name of medium or (in other arenas) of operator. The trepidation arises from the fact that these people, who should only be channeling but who of course possess their own identities and aims, have the potential to enable but also to interfere with communication: to deliver the message but also to deform its basic structure, and thus its basic import, from the (real or only apparent) inside of the channel. Daniel Dunglas Home deforms communication by falsifying authorship. Telegraphers and others might do so by multiplying reception—by reading what they should only be transmitting, and perhaps sharing message contents with others. Latimer's narration—exceptionally palpable as a relay to a reader, but perverting that relay because unreliable—emblematizes in the realm of literature the intricacies and even treacheries at risk for a culture that increasingly entrusts its messages to human media.

Fanny Brand's Electrical Powers

If the literary medium within the plot of Rudyard Kipling's "Wireless" is for his part fairly reliable, that is because he is such a clear echo of the person he channels. Mr. Shaynor's role in the story is a relative creature's—relative to a great poet, that is—which is only fitting in a work that accentuates his feminization. As its title suggests, "Wireless" concerns Guglielmo Marconi's radio elaboration on the electric telegraph, which had been introduced only recently to the British public. A young electrician named Mr. Cashell is trying from the back of his uncle's drugstore to receive a message from a station at Poole. The unnamed narrator, who has come out just to observe this telegraphic trial, is eventually distracted by a second kind of transmission: the store's assistant apothecary is automatically writing the lines (in rough form) of John Keats's "The Eve of St. Agnes." How to explain the poetically ignorant Mr. Shaynor's composition? The narrator theorizes that because Shaynor works the job he does, suffers from the first stages of tuberculosis, and has a "lady-friend" named Fanny, he corresponds enough to the druggist, consumptive, and lovesick Keats to be able, as it were, to recall him into being.[27] Cashell's attempted

27. Rudyard Kipling, "Wireless," in *Traffics and Discoveries* (1904; New York: Doubleday, Page, 1914), 201. All further page references to "Wireless" are parenthetical within the text. Kipling published the story first in a periodical in 1902, then in *Traffics and Discoveries* two years later.

induced electrical current is matched by the spiritualistic phenomenon of an "induced Keats" (213).

In one respect "Wireless" is about literary production, dramatically suggesting the reality of aesthetic inspiration.[28] But the most provocative features of the story have less to do with questions of artistic origin than with electricity. Victorian and turn-of-the-century identifications of electricity's effects with the paranormal hint at perceptions of its oddity. For all its tangible results, it seemed, as one writer stated it, "shadowy, mysterious, and impalpable. It still lives in the skies, and seems to connect the spiritual and the material."[29] In a comic turn, Oliver Wendell Holmes's much reprinted 1890 poem "The Broomstick Train, or the Return of the Witches" represents the electricity that powers trolley cars as an "evil-minded witch" who "will do a mischief if she can," but whose feminine magic ultimately succumbs to the masculine authority of the switchman.[30] Other feminizations of electricity drew on that similarly exciting yet enigmatic force, women's sexuality, as when the 1900 World Exposition enticed visitors to its electricity palace with an image of a nude goddess perched atop a water castle.[31]

Around 1900, women's increasing adoption of jobs and other positions outside the home made their bodies all the more alluring because they were now more available to the public eye. The rhetoric of anti-domesticity and sexual unrestraint around the New Woman typifies a time when the female sex seemed perilously unconfinable and unpredictable. This context makes particularly resonant what I take to be a crucial motif in Kipling's story, the conjoint fugitiveness of women and electricity. Electricity in "Wireless" is familiarly feminine, occult, and seductive; but unlike in Holmes's vision, it escapes male control, and indeed, the story shows men enthralled by its

28. See J. M. S. Tompkins, *The Art of Rudyard Kipling* (London: Methuen, 1959), 90–94; Elliot L. Gilbert, *The Good Kipling: Studies in the Short Story* ([Athens] Ohio University Press, 1970), 164–66; Sandra Kemp, *Kipling's Hidden Narratives* (Oxford: Blackwell, 1988), 35–38; Martin Seymour-Smith, *Rudyard Kipling* (New York: St. Martin's Press, 1989), 321–25; Helen Pike Bauer, *Rudyard Kipling: A Study of the Short Fiction* (New York: Twayne, 1994), 97–99; and Roger Luckhurst, *The Invention of Telepathy: 1870–1901* (Oxford: Oxford University Press, 2002), 178–79.

29. Quoted in Daniel J. Czitrom, *Media and the American Mind: From Morse to McLuhan* (Chapel Hill: University of North Carolina Press, 1982), 9.

30. On Holmes's poem, see David E. Nye, *Electrifying America: Social Meanings of a New Technology, 1880–1940* (Cambridge, MA: MIT Press, 1990), 150–52.

31. Christoph Asendorf, *Batteries of Life: On the History of Things and Their Perception in Modernity*, trans. Don Renau (Berkeley: University of California Press, 1993), 164. As Julie Wosk notes, Villiers de l'Isle Adam's *L'Eve future* (1886) continues long established associations between electricity and eroticized, sometimes siren-like femininity, in that its electrified automaton replicates a fallen woman whose beauty is treacherously alluring; *Women and the Machine: Representations from the Spinning Wheel to the Electronic Age* (Baltimore: Johns Hopkins University Press, 2001), 79–80.

powers. In its seeming evanescence, electricity becomes a symbol of female restiveness and the consequent affront to gender roles within the contemporary social landscape. Recalling the feminization of spirit channeling gives us a first clue to the gender unease in the tale's portrait of Shaynor's entrancement. As it would happen, Kipling had personal acquaintance with feminine mediumship through his nervously fragile sister Alice, or "Trix" (who was, incidentally, one of the four automatists involved in the case of the cross-correspondences); like Shaynor, Trix sometimes used automatic writing to channel poetry.[32]

The initial cause of the apothecary's trance is not, as the narrator assumes, the intoxicating drink he concocts for Shaynor, but rather the sight of Fanny Brand, who comes to the store to ask Shaynor out for a walk. When she first enters—before he has accepted the narrator's concoction—Shaynor's physical demeanor immediately changes and his eyes shine "like a drugged moth's" (201). Even after her exit from the narrative, she continues to exert a spell over him by way of a toilet-water advertisement featuring the "seductive shape" of a pearl-adorned young woman who resembles her. The ad acts for Shaynor like a "shrine" (204); surrounded by the incense of pastilles, it draws his gaze throughout his trance and seems to produce and sustain it. "At each page," the narrator says of the letter-writing that just precedes Shaynor's loss of consciousness, "he turned toward the toilet-water lady of the advertisement and devoured her with over-luminous eyes," looking the very "incarnation of a drugged moth" (205–6).

Fanny embodies a freedom, especially sexual freedom, that upends traditional romantic relations, as well as imitates electricity's own elusiveness. It is she who plays suitor in her interactions with Shaynor, leaning "confidently across the counter" and responding peremptorily to his objections to a stroll, at last turning her flirtations on the narrator to make her case: "You['ll] take the shop for half an hour—to oblige *me*, won't you?" (201). Her assured demeanor opposes and complements Shaynor's blushing and his status as domestic courtship object. The store he works in appears to have lost all tone of masculine trade, reverting to a site of leisure and languor. The owner's nephew has taken over the "house" for his personal experiments (197, 200, 202); the owner himself lies sick in a bedroom above; and his tubercular assistant plans to pass the evening warmed by a stove and blanket, writing a letter—probably a love letter, given his frequent glances at the advertisement as he writes and the bunch of perfumed, scalloped-edged missives he

32. Trix's poetic transmissions reportedly originated with her ancestors. Ina Taylor, *Victorian Sisters* (London: Weidenfeld and Nicolson, 1987), 146–47, 171–72.

peruses before beginning. These homey trappings dissociate Shaynor from his professional position and the prerogatives of male identity it implies, including the prerogative to move and act at liberty. He remains fairly rooted to old Cashell's place, returning from his walk after only a few pages. By contrast, Fanny's movements are ambiguous; her presence within the store is fleeting, nor does the reader know where she disappears to once Shaynor comes back unaccompanied. Explaining Shaynor's poetic "induction," the narrator pinpoints Fanny as "key," saying that she "approximately represents the latitude and longitude of Fanny Brawne" (213). His language accentuates the idea of her spatial position, but also its "approximate" quality, intimating the fitful uncertainness of her whereabouts. Her advertisement double, which obviously accounts for her surname, underscores her roving, because as a mass-produced image it has presumably been displayed in any number of locations.

Further, through this image "Wireless" allies Fanny with electricity, an entity whose similarly invisible actions through space awe young Cashell with their "strangeness" (210). For what is so riveting to Shaynor about the toilet-water advertisement is not just its resemblance to his girl but also the play of brilliant color that spills onto it—a special effect of electrical lighting. Old Cashell has filled his place with lights, and the ones set in front of three colored glass jars in the store's window throw "inward three monstrous daubs of red, blue, and green" (205). One of these daubs happens to be shining on the advertisement, causing Shaynor, "drugged moth" that he is, to stare all the more fixedly at the pictured woman; her physical assets are "unholily heightened by the glare from the red bottle in the window" (204). The fact that gazing at this spectacle sends Shaynor into a mediumistic trance reconfigures in an ironic way the modern spiritualistic principle that electrical forces are what power communications with the dead.

The illuminated advertisement makes electricity and sexualized womanhood inextricable, as they are in "Wireless" generally. Shaynor tells the narrator about the mishap that occurred during young Cashell's last experiment: the pole he had attached to the rooftop of a local hotel electrified its water supply so that "the ladies got shocks when they took their baths" (197). The anecdote calls up a picture, as Martin Seymour-Smith vividly puts it, of "'electrified' naked women."[33] Just as the ladies in the hotel were bathed in electricity, Fanny's advertisement lookalike "bathe[s] in the light from the red jar" (210); yet while in the first case electricity renders the female body

---

33. Seymour-Smith, *Rudyard Kipling,* 322.

helpless, in the second it is Shaynor's own body that becomes so. Handed a second glass of the narrator's concoction, Shaynor does not even appear to sip from it—instead the drink reminds him once again of his radiant "shrine":

> "It looks," he said, suddenly, "it looks—those bubbles—like a string of pearls winking at you—rather like the pearls round that young lady's neck." He turned again to the advertisement where the female in the dove-coloured corset had seen fit to put on all her pearls before she cleaned her teeth.
>   "Not bad, is it?" I said.
>   "Eh?"
>   He rolled his eyes heavily full on me, and, as I stared, I beheld all meaning and consciousness die out of the swiftly dilating pupils. His figure lost its stark rigidity, softened into the chair, and, chin on chest, hands dropped before him, he rested open-eyed, absolutely still. (206–7)

In a feminization of his desire, Shaynor is made flaccid by lust—his for Fanny Brand, parallel with Keats's for Fanny Brawne. Later the narrator watches Shaynor automatically write several lines of "bald prose—the naked soul's confession of its physical yearning for its beloved," which are interpreted as Keats's "raw material" for "The Eve of St. Agnes" (214). In the final astounding moments of Shaynor's transcription—when he is gripped by a literally illuminating "Power," as if electricity itself—he experiences a climax that is positively orgasmic.

> Here again his face grew peaked and anxious with that sense of loss I had first seen when the Power snatched him. But this time the agony was tenfold keener.... It lighted his face from within till I thought the visibly scourged soul must leap forth naked between his jaws, unable to endure....
>   "Not yet—not yet," he muttered, "wait a minute. *Please* wait a minute. I shall get it then—... *Ouh,* my God!"
>   From head to heel he shook—shook from the marrow of his bones outwards—then leaped to his feet with raised arms, and slid the chair screeching across the tiled floor where it struck the drawers behind and fell with a jar. (217–18)

Shaynor's trance-state ecstasy mirrors the sexual enjoyments sketched in "The Eve of St. Agnes" itself. The poem describes Porphyro's infiltration of his sweetheart Madeline's bedroom, where, concealed from sight, he watches as she undresses, while her own thoughts are absorbed by the legend of St. Agnes's eve, on which it is said dutiful

> Young virgins might have visions of delight,
> And soft adorings from their loves receive
> Upon the honey'd middle of the night.[34]

Once in bed Madeline slips more and more into a "sort of wakeful swoon" that turns gradually into a "stedfast spell," a charm cast by her erotic imaginings of Porphyro (ll. 236, 287). His presence in her room at last rouses her, but so indistinguishable are her slumbering visions from his sexual advances that she confuses wakefulness with sleeping, mistaking the moment of consummation for the continuation of her fantasy:

> Into her dream he melted, as the rose
> Blendeth its odour with the violet,—
> Solution sweet. (ll. 320–22)

Clearly this tale of carnal enchantment allegorically grounds "Wireless," yet whereas the former pairs male seducer with female somnambulist, the latter transposes genders, aligning Shaynor with Madeline. In a story that pays oblique but faithful homage to Keats's poem (Fanny wants to walk out by the church of St. Agnes; the store's electric lights create the illusion of "slabs of porphyry" [205]; Shaynor's red, black, and yellow blanket is identical in color to the tablecloth spread out in Madeline's chamber; and so on), this conspicuous switching of elements emphasizes the unmanliness of Shaynor's psychical ravishment, while also attributing a peculiarly virile influence to Fanny.

Electricity not only augments the effect of Fanny's image on Shaynor but in a sense spellbinds young Cashell too. Cashell explains Hertzian electromagnetism to the narrator with evident pride, "as though himself responsible for the wonder" (203). Another demonstration finishes with his proclaiming, "There's nothing we sha'n't be able to do in ten years" (209). Yet his cocksure attitude is tempered by his reverence for electricity as a "magic" about which "nobody knows" the real essence (202). He speaks in deferential capitals of "Electricity," "It," and the "Powers" (202–3) and reflects on the restlessness of the force. "*That's* the Power—our unknown Power—kicking and fighting

---

34. John Keats, "The Eve of St. Agnes," in *The Longman Anthology of British Literature,* ed. David Damrosch and Kevin J. H. Dettmar, 3rd edition, vol. 2, 935–46 (New York: Longman, 2006), ll. 47–49. All further line number references to "The Eve of St. Agnes" are parenthetical within the text.

to be let loose," he announces as sparks fly from his apparatus. "There she goes—kick—kick—kick into space." Minutes later the narrator hears a "*kiss—kiss—kiss*" from where the halyards on the rooftop are brushing up against Cashell's installation pole (210). This erotically uncontainable "she"-"Power" appears continuous with that which pulls Shaynor into a trance: "his face grew peaked and anxious with that sense of loss I had first seen when the Power snatched him. But this time the agony was tenfold keener. As I watched it mounted like mercury in the tube" (217–18). Not coincidentally, that simile recalls the narrator's description of Cashell's coherer, the tool that senses Hertzian waves, as "a glass tube not much thicker than a thermometer" (202), fashioning Shaynor as a personification of Cashell's receptive device. When the narrator tries to chat with him, Cashell's reaction conveys that he has in his own way given himself over to the mysterious "Power": "a frail coil of wire held all his attention, and he had no word for me bewildered among the batteries and rods" (202).

Shaynor's embodiment of the telegraphic coherer emphasizes the effeminacy of his relay. It was Oliver Lodge, a physicist as well as spiritualist, who invented the coherer, later perfected by Marconi for wireless transmission. Lodge's achievement involved reworking the "Branly tube," a test tube containing dispersed metal filings that stuck together when hit by an electric charge, enabling a current to pass through; as he showed, the cohered metal filings could detect Hertzian waves, or electromagnetic force traveling through space.[35] To orient the narrative picture of automatic writing around the coherer is to stress the feminine properties that made spiritualism work: the medium's sensitivity and receptivity were cohering forces, bringing together that which would otherwise be dispersed, the living and the dead. Shaynor epitomizes this instrumental turn to the other; he is as subtly responsive, and as much of a vessel, as Lodge's Branly tube.

Such insinuations of his emasculation complement the tale's sustained interlinking of electricity, rapt states, and female sexuality. Its vision of electricity as a mastering force intrigues, too, by its contrast with familiar tales of masculine triumph around technological invention. With its captivated men, "Wireless" highlights the air of baffling potency around electricity, suggesting a less assured ability to guide the course of modernity. Those who try to exploit the new "Power" sometimes miss their mark: the last scene has the younger Cashell picking up a botched wireless transmission between two ships at sea, a failure to contact that one of the senders calls

---

35. See Gavin Weightman, *Signor Marconi's Magic Box: The Most Remarkable Invention of the 19th Century & the Amateur Inventor Whose Genius Sparked a Revolution* (Cambridge, MA: Da Capo, 2003), 17–18 on Lodge's coherer and 91–92, 94–97, and 286 on his fascination with spiritualism.

"*Disheartening—most disheartening*" (220).[36] Above all, the story is a tableau of the perplexities of modern gender. If the druggist Shaynor's somewhat lowly social status makes it difficult at first to see him and his girlfriend as clear representatives of contemporary challenges to bourgeois morality, of the kind the New Woman exemplified, it is still more difficult to ignore the repeated implications of his womanliness vis-à-vis Fanny's dominance and of electrified female irrepressibility. These make "Wireless" appropriate to an age in which women's wider field of action seemed to be confounding the meanings of masculinity and femininity altogether.

Gender Negotiations and the Modernist Channel

Shaynor is ravished by the process of poetic mediumship. George Eliot's Latimer only wishes he could be so inspired (he initially mistakes his vision of Prague for a poet's "spontaneous creation," the sign of a "newly liberated genius" [64]), but his insights, whether real or imagined, also presume a body and psyche liable to incursions. Opposed to these texts is a theory of poetic channeling published within a few decades of "Wireless," T. S. Eliot's "Tradition and the Individual Talent" (1919). In this seminal essay, Eliot makes the "surrender" of self key to good poetry-writing: "the poet has, not a 'personality' to express, but a particular medium, which is only a medium and not a personality, in which impressions and experiences combine in peculiar and unexpected ways."[37] Subduing personality assists the poet in conveying the voices of the departed: "we shall often find that not only the best, but the most individual parts of his work may be those in which the dead poets, his ancestors, assert their immortality most vigorously" (38). T. S. Eliot's picture of impersonal, haunted writing echoes séance discourse.[38] Yet it also essentially diverges from this discourse in stressing the poet-channel's self-security and intelligent volition.

---

36. Jeffrey Sconce links this moment to early twentieth-century ideas about the seemingly aimless and isolating communications of wireless radio; *Haunted Media: Electronic Presence from Telegraphy to Television* (Durham: Duke University Press, 2000), 69–70. Also on the story's ties to modern science, technology, and communication, see Gillian Beer, "'Wireless': Popular Physics, Radio and Modernism," in *Cultural Babbage: Technology, Time and Invention,* ed. Francis Spufford and Jenny Uglow, 149–66 (London: Faber and Faber, 1996); and Peters, *Speaking into the Air: A History of the Idea of Communication* (Chicago: University of Chicago Press, 1999), 106.

37. T. S. Eliot, "Tradition and the Individual Talent," in *Selected Prose of T. S. Eliot,* ed. Frank Kermode (London: Faber and Faber, 1975) 40, 42. All further page references to "Tradition and the Individual Talent" are parenthetical within the text.

38. Helen Sword remarks the spiritualistic influence; *Ghostwriting Modernism* (Ithaca: Cornell University Press, 2002), 94.

led many spirits to laud them as well tuned musical instruments—while male mediums faced accusations of homosexuality.[39]

In a sense, early twentieth-century mediums were more discernibly mentally invested than their precursors. Some saw their vocation as an *"intellectual adventure"* because they delivered automatic writings in classical languages or concerning ancient cultures.[40] In addition, the channel's own mind was discussed as an important basis of translation from the afterworld to our world. The automatic writer expressed or "interpreted" spiritual concepts by way of her own vocabulary; in one medium's view, "the brain and memory of the medium or interpreter are used to a greater extent than the average observer imagines."[41] Sometimes the medium might even be more or less conscious while automatic writing, able to see and comment on her production or not perceiving the actions of her hand as involuntary.[42] There were inherent limits in these scenarios, though. For mediums also affirmed their prior personal ignorance of the scholarly subjects about which they communicated; more fundamentally, they habitually denied credit for their automatic writings, portraying themselves as stenographic instruments of the spirits.[43] As Bette London concludes, twentieth-century automatic writing allowed for women's rare involvement in intellectual matters and rested on a profoundly ambiguous model of collaboration in which, as much as the "spirit" or the séance sitters, the "medium's own mind (... its different levels of consciousness)," shared a stake in the authorial process.[44] Yet it bears recalling the persistent feminization of the writing position that, along with the ghostly itself, disorders authorial presence here. The premise of the female

39. On the boost to spiritualism around wartimes and on the comparisons of mediums to musical instruments, see Sword, *Ghostwriting Modernism,* 46–47 and 20 respectively. On male mediums' purported homosexuality, see Jenny Hazelgrove, *Spiritualism and British Society Between the Wars* (Manchester: Manchester University Press, 2000), 5.

40. Bette London, *Writing Double: Women's Literary Partnerships* (Ithaca: Cornell University Press, 1999), 132. Sword likewise discusses the "intellectual," in that case literary, pursuits of mediums of this period (13–14, 39–48).

41. Quoted in London, *Writing Double,* 166.

42. Ibid., 192–94.

43. On mediums' affirmed ignorance of learned subjects, see ibid., 133–35, and on their repudiation of authorship, 163–65.

44. Ibid., 195. See also London, "Mediumship, Automatism, and Modernist Authorship," in *Gender in Modernism: New Geographies, Complex Intersections,* ed. Bonnie Kime Scott, 623–32 (Urbana: University of Illinois Press, 2007). Perhaps London's most sensational example of collaborative mediumistic authorship is the medium Geraldine Cummins's successful copyright lawsuit in 1926 against a sitter who was also the proofreader and transcriber of some of her automatic writings (*Writing Double,* chap. 5 passim). But even in this case, as London says, the medium was seeking a legal title it was her vocation to renounce; in the very book she obtained copyright over, Cummins is credited as "recorder" rather than author (*Writing Double,* 153).

medium's unconscious wisdom opposes the signature T. S. Eliot's poet puts to his erudite work—while at the same time, the emphasis on her own mind as a crucial substrate modernizes mediumship, combining womanly receptivity with a form of personal mental consequence.

The inspired feminine sensitive figures within the walls of Modernism itself. While Gertrude Stein investigated automatic writing as a matter of mundane physiology, W. B. Yeats was intrigued by it as an otherworldly phenomenon seemingly embodied in his wife, George, whose automatic writings supplied many of the ideas that would appear in his mystical treatise *A Vision* (1925, 1937).[45] In his portrait of the seer Tiresias in *The Waste Land* (1922), T. S. Eliot borrows once again from a mediumistic tradition, but here in a more wholesale way that also borrows its presumptions about gendered sensitivity. "The Fire Sermon" section of the poem gives us Tiresias prophesying the "bored and tired" typist's sexual encounter with the rapacious "young man carbuncular."[46] The scene doubles the seer's self-division, his/her "throbbing between two lives" (l. 218): there is not only here an internal bi-genderedness—Tiresias has lived both male and female lives—but also an interdependence of his/her story and identity with the typist's. Tiresias repeatedly calls attention to this interdependence—"I too awaited the expected guest" (l. 230); "And I Tiresias have foresuffered all / Enacted on this same divan or bed" (ll. 243–44)—as if, like so much of the rest of the poem, these two characters composed a unit in pieces. In Tiresias, Eliot hearkens back to a Victorian, and especially a Victorian spiritualistic, point of view that links both spirituality and empathic receptivity to femininity. That is to say, the seer's female physique and personal history create his/her moral capacity to feel with another—the capacity not only to foresee but also to "foresuffer" the miseries of the typist's sexual coupling with her odious visitor. While the typist herself remains almost unconscious of his gross affect   afterward she is "Hardly aware of her departed lover" and mech?   es about her room again, alone," "smoothes her hair with automat.   And puts a record on the gramophone" (ll. 250, 254–56)—Tiresias's   ary sympathy conveys all the emotion of the scene, glossing it for the lo. it represents.

---

45. For an excellent analysis of the complexities of the Yeatses' literary collaboration, see London, *Writing Double,* chap. 6. On Stein see Barbara Will, "Gertrude Stein, Automatic Writing and the Mechanics of Genius," *Forum for Modern Language Studies* 37 (2001): 169–75. As Sword argues, even those Modernists who ridiculed spiritualism may have been drawn to its themes and formal and linguistic properties, including its abrupt transitions between different voices and perspectives.

46. T. S. Eliot, *The Waste Land,* in *The Longman Anthology of British Literature,* ed. David Damrosch and Kevin J. H. Dettmar, 3rd ed., vol. 2, 2518–31 (New York: Longman, 2003), ll. 236, 231. All further line number references to *The Waste Land* are parenthetical within the text.

Reading the emblem of Eliot's sexually casual typist to its fullest—registering its tragedy—necessitates recognizing the fissure in the typist-Tiresias pairing: the modern woman's estrangement from her natural reserves of moral and mystical sensitivity.

The conjoint experience of the oblivious typist and compassionate Tiresias suggests the ongoing correlation of automatism with sympathy, as well as the multiple forms it might take depending on the conceptual pressures of the representation. The notion of feminine occult feelingness extends beyond the nineteenth century even as the cultural situations it responds to change over time. Eliot's characterization of the seer is reminiscent of the Victorians, but it is also Modernist, in the sense that it speaks to a particularly Modernist longing for ethical and spiritual plenitude. While "Tradition and the Individual Talent" conveys Eliot's attempt to safeguard individuality, *The Waste Land* laments individuality in its starkest form, because it serves as a sign of spiritual drought, of a lack of a collective moral order in the twentieth century. Hence the masculinizing of mediumship in one work gives way in another to an older, more comforting model. The typist in the *Waste Land,* like electricity in "Wireless," is the technological symbol of a disoriented modernity. The poem's turn toward affective femininity hints at a belief in altruistic womanhood as the essence of social interconnection, a perspective vital to visions of communication relays before Eliot's day and long after.

Epilogue

More than a century ago, female-tended relays helped people to communicate across space and time. Thus arose a certain image of women's relationship to information and dialogue, an image sketched and resketched in the stories this book has examined. The women in these tales are media that transmit, connect, record, and detect; they are feeling bodies with minds that fracture or dissociate to become routes to other selves. Now, at the turn of the twenty-first century, the possibilities for at-a-distance contact have changed dramatically. Yet these changes are far from wholesale, and older ideas of communications and of the women enabling them continue to haunt the present, adapted for new purposes.

"Alison was a Sensitive: which is to say, her senses were arranged in a different way from the senses of most people," Hilary Mantel tells us at the beginning of her tale of an English spirit medium, *Beyond Black* (2005). At the psychic extravaganza Alison attends, all the other mediums are also female, except for two "lugubrious and neglected" men, Merlin and Merlyn. Years after the rappings at the Fox house brought to the popular mind Morse's telegraph, mediumship still involves a technology of the spirit, though one perhaps more compatible with telegraphs than latter-day machines. "So I think it's my electromagnetic field": this is how Alison interprets the malfunction of her computer and tape recorder; "I think it's hostile to modern technology."

Mediums themselves still sometimes become machines: when one loses all self-control onstage, another declares, "She's gone on automatic."[1] But more essential to Alison's work, as Mantel's exposition implies, is sensitivity. When Princess Diana suddenly dies, Alison feels her transition to the afterlife so powerfully that it sickens her. Her work involves constant emotional turmoil, as she identifies the band of malevolent spirits who besiege her with the men her mother allowed to molest her as a child. Alison must literally keep these demons at bay, and the reader is left to contemplate all it means to call them the conjurations of her psyche. To what extent are these real spirits and to what extent evidence of a plagued, disordered mind?

A traveling show-woman, Alison knows to pander to local audience expectations and that "the farther north you go, the more the psychics' outfits tend to suggest hot Mediterranean blood, or the mysterious East."[2] The face of the Psychic Readers Network (which made news when the Federal Trade Commission charged it with fraud in 2002) was one Miss Cleo, whose Jamaican accent reminded modern audiences of the special occult powers of "exotics."[3] Today psychic contact is mediated through hotline operators; it is also on full display, whether in fictional or "reality" stories, on television. The latest personality to emerge in a genre made notable by John Edward, of *Crossing Over with John Edward* fame, is the British medium Lisa Williams, whose shows on Lifetime Television featured her chatting with strangers and giving them messages from the dead. And as of this writing, not one but two television series revolve around women spirit mediums—NBC's (soon to be CBS's) *Medium* and CBS's *Ghost Whisperer*. In the first, Allison DuBois is a detective device of the first order, combining her role as a homemaker with her work as a crime-solving consultant for the district attorney and other investigators. *Ghost Whisperer* takes a more time-worn narrative tack, with the medium helping to pacify spirits with lingering business in the mortal world. As the blurb on CBS's website describes, Melinda Gordon's husband "worries about the emotional toll this work is taking on his wife as they embark on a new life together."[4] Both shows offer mediumship as a vocation that must be juggled with domestic obligations; *Ghost Whisperer* especially implies

---

1. Hilary Mantel, *Beyond Black* (New York: Picador, 2005), 7, 72, 295, 338.

2. Ibid., 163.

3. Federal Trade Commission, "FTC Charges 'Miss Cleo' Promoters with Deceptive Advertising, Billing and Collection Practices," http://www.ftc.gov/opa/2002/02/accessresource.htm.

4. CBS Corporation, "About Ghost Whisperer," CBS Shows: Ghost Whisperer, http://www.cbs.com/primetime/ghost_whisperer/about/.

that it siphons off a store of affective commitment that would otherwise be given over to the home.[5]

My turn here to the non-literary continues this book's consideration of female mediation at multiple cultural sites, not excluding popular fictional forms. Like Marie Corelli's and George Du Maurier's novels, television serials indicate the wide scope of conceptions of gendered relays. Corelli and Du Maurier are also important touchstones in suggesting the generic malleability of the place of the female channel. With their tales of space travel and interplanetary angels (or angelic aliens), *A Romance of Two Worlds* and *The Martian* are strange hybrids between spiritualistic narrative and science fiction, reminding us that both genres involve the probing of a great beyond. In the twentieth century, the theme of intergalactic investigation is perhaps nowhere more pronounced than in the *Star Trek* franchise: the starship USS *Enterprise*'s "mission" is to "explore strange new worlds" and to "boldly go where no man has gone before." If *Star Trek* has always been anchored by the idea of exploration, of gleaning knowledge about other planets and cultures, it has also often imagined women as media of that knowledge. In the original 1960s series, the only woman officer on the bridge is Lieutenant Nyota Uhura, chief communications officer, in charge of the control panel the starship uses to dialogue with other ships.[6] The scene is different but still familiar in *Star Trek: The Next Generation,* the installment of the franchise that ran three decades later. Now there are two women serving as permanent officers on the bridge, but only one of them, Counselor Deanna Troi, will remain after the first season. Troi mimics Uhura's job of mediating exchanges between the crew and alien races, yet her relays work not through a control panel but rather through her inherited abilities as an "empath." Repeatedly the camera zooms in on her face to show it awash with received emotions. Just as the nineteenth-century medium helped to constitute an electrically organized séance "battery," Troi's powers form part of a larger communication machinery. Often we see her seated before the bridge's main audiovisual link, directly beside the ship's captain, poised to share her intuitions about the figures onscreen for diplomatic or strategic purposes: in other words, to publicize the personal, to make private states of mind a matter of political scrutiny.

---

5. In contrast to these two current series are two others, CBS's *The Mentalist* and USA's *Psych,* each of which centers on a character who only seems psychically sensitive due to excellent observational and deductive skills. In both cases, interestingly, these characters are male.

6. Here's the gist of her character as encapsulated on the popular website Allscifi.com: "Uhura was the most senior woman officer on the show, and she was little more than a glorified telephone operator!" "Uhura as the Barbie Doll Moesha of the 1960's," http://www.allscifi.com/Board.asp?BoardID=196.

In *Galaxy Quest* (1999), a film spoof about actors from a defunct *Star Trek*–like series who end up real outer-space heroes, the woman who plays the communications officer simply repeats the transmissions spit out by the ship's computer.[7] On one level, this parodies the notion that mediating women are automatized extensions of the machines they work. On another level, it suggests the redundancy, as technologies become increasingly sophisticated, of mediating women altogether. That kind of technological development is also evident within the *Star Trek* franchise, in the installment titled *Star Trek: Enterprise*. This gives us yet another female communications officer, Ensign Hoshi Sato, whose main job is the translation of alien tongues. But *Enterprise* is a prequel to the original *Star Trek* series; by the time (in diegetic terms) of *Next Generation,* set two centuries later, where Sato once served there is now something called the universal translator.

This fictional trend echoes our lived experience of communication technology, particularly the computer. As Darren Wershler-Henry argues, the continued usage of the QWERTY keyboard may give the erroneous impression that the computer is simply a souped-up typewriter; in reality the two are fundamentally different technologies, with different methods for input, output, and processing. Most centrally, the typewriter is "profoundly non-networkable" with other machines.[8] Computing, by contrast, functions through networks and software, and its speed and the user-friendliness of its interface make for general ease of operation. All of this means a diminishment of the role of the human medium, in business settings most notably: executives can, when necessary, do the work of preparing and sending their own documents without the aid of a secretary.[9]

But diminishment is not obsolescence. Secretaries—and administrative assistants, as many office workers are called today—still process information and handle contacts between the boss and his/her clients or business partners. Moreover, this is still largely the work of women.[10] But along with changes in information technologies have come elaborations in women's office jobs. Administrative assistants combine traditional secretarial duties like scheduling and document production with duties formerly reserved for managers and

---

7. Thanks to Michael North for calling my attention to this character.

8. Darren Wershler-Henry, *The Iron Whim: A Fragmented History of Typewriting* (Ithaca: Cornell University Press, 2005), 274; see 260–74 on distinctions between the typewriter and the computer.

9. On the decline of certain mediatory functions associated with femininity and the body, see Douglas A. Brooks, "Technology, Embodiment, and Secretarial Mediation," in *Literary Secretaries Secretarial Culture,* ed. Leah Price and Pamela Thurschwell, 129–50 (Aldershot, UK: Ashgate, 200

10. See for example Blanche Ettinger, *Opportunities in Administrative Assistant Careers* (New McGraw-Hill, 2007), 53, table 4.5.

professionals, such as crafting presentations and writing reports.[11] In the context of these expanded responsibilities, there is little place for the notion of distracted or self-forgetting femininity. Indeed, that notion of the woman in the office was beginning to be pressured as early as the turn of the twentieth century, when the private secretary was distinguished from the mere typist by, besides a more personalized relationship with the boss, her self-application and -motivation.[12] Novels like Marie Corelli's and Bram Stoker's, not to mention the discourse around séance mediumship, indicate how assailable the premise of female automatism was even at that time. Now, given the modern office's technological complexity, a worker unable to contribute her wits might be as extraneous as *Galaxy Quest*'s communications officer.

The business office has altered over the years, with machines replacing human functions; yet a handbook on administrative assistant careers implies there are limits to this supplanting: "many secretarial and administrative duties are of a personal, interactive nature and, therefore, they cannot be easily automated," which ensures the ongoing importance of positions in the field. Among other things, these positions require "excellent tact and communication skills." We are back to an awareness of the feeling subtext of office exchanges: like the private secretary of a century ago, the twenty-first-century administrative assistant demonstrates a perceptiveness about others' emotions (here called "tact"), and one of the six most important capacities within the profession is "intuition." Looked at more closely, even self-motivation, another ongoing feature of the job, derives from something like this quality. If the administrative assistant must be able to "take the initiative and not wait to be asked to do something," one imagines that figuring out what an employer wants done without being told demands an acute sensitivity to his/her will.[13]

Moreover, the closeness of the relationship thus developed implicitly helps to secure any information the boss may share. As another handbook puts it, "Trust is another factor in a successful relationship with an executive. Many persons are under a great deal of pressure and will use you as a sounding board for confidential maters... Absolute loyalty to your supervisor is therefore essential and can mean the difference between working *with* someone instead of simply *for* someone."[14] There is an emphasis here on harmony that aggrandizes administrative assistant work even while it subtly indicates the sympathetic talents the work may require. Where companies once valued

---

11. Ibid., 15–17, 37–41; U.S. Department of Labor, "Secretaries and Administrative Assistants," J.S. Bureau of Labor Statistics, http://www.bls.gov/oco/ocos151.htm.

12. Angel Kwolek-Folland, *Engendering Business: Men and Women in the Corporate Office, 1870–1930* (Baltimore: Johns Hopkins University Press, 1994), 59.

. Ettinger, *Opportunities*, 99–100, 7, 28.

The New Office Professional's Handbook, 4th edition (Boston: Houghton Mifflin, 2001), 67.

women's supposed mental weakness, here established affinities help to protect the privacy of office communications.

The importance of an affective investment in the work of information management and communication is even more apparent in modern call centers. Often the main purpose of these centers is to field service calls for businesses. Internationally, an average of 69 percent of the workforce is female,[15] and researchers suggest a key reason involves assumptions about women's empathetic capacities. Managers speak of the need for a "caring attitude," and women are thought to be "particularly skilled at listening to and empathising with customers" due to natural differences between the sexes.[16] In the course of one study, employment screeners picked as one of their center's two best workers a woman who came across on the phone as a "warm person." A remarkable demonstration of her talents occurred when she took a call from a customer who, in the process of seeking out financial information about her recently deceased husband, began discussing her despondency. The worker reacted with a "kind and gentle" manner and assured "the customer to place her faith in God, that he would help her through this trying time." Though the employee was "relatively leisurely" in her delivery—and significantly slower than the other best worker at the center, a briskly efficient man—the liability to call-turnover time did not impede screeners' appreciation of her style.[17] In fact, the authors of the study conjecture, screeners slightly favored a "supportive" demeanor over efficiency, and this, coupled with gender stereotypes about the femininity of this type of behavior, accounts for the greater number of women hired.[18]

Call-center work privileges other-focused attunement, even a quasi-intimacy. The particular gendering of this form of emotional labor becomes clearer when we look at the study of another center.[19] As a way of ensuring

---

15. David Holman, Rosemary Batt, and Ursula Holtgrewe, "The Global Call Center Report: International Perspectives on Management and Employment," DigitalCommons@ILR, Cornell University ILR School, http://digitalcommons.ilr.cornell.edu/cgi/viewcontent.cgi?article=1012&context=reports, 9. For a general picture of scholarship on call centers, see Bob Russell, "Call Centres: A Decade of Research," *International Journal of Management Reviews* 10 (2008): 195–219.

16. Vicki Belt, Ranald Richardson, and Juliet Webster, "Women, Social Skill and Interactive Service Work in Telephone Call Centers," *New Technology, Work and Employment* 17 (2002): 25, 26.

17. Roberto M. Fernandez and M. Lourdes Sosa, "Gendering the Job: Networks and Recruitment at a Call Center," *American Journal of Sociology* 111 (2005): 894.

18. Ibid., 895. The converse of this is that male workers, often concentrated in the (more remunerative) technical-help segment of the call center, may be expected to lack social communication skills. See Belt, Richardson, and Webster, "Women, Social Skill," 25–26.

19. Here I use emotional labor in the sense articulated in Arlie Russell Hochschild's seminal *The Managed Heart: Commercialization of Human Feeling*, Twentieth Anniversary Edition (1983; Berkeley: University of California Press, 2003). Hochschild writes of emotion management as the work of producing socially appropriate emotions and emotional displays, while emotional labor denotes the

company loyalty and meeting internal goals, both male and female workers were pushed by supervisors to establish subtly sexualized ties with callers of the opposite sex. But the form of these ties was dramatically different for the two sexes: whereas the men engaged in "banter"—telling stories and jokes, bragging and calling attention to their attractiveness—and saw this as a fun part of the job, the women concentrated on "'really getting to know' a particular client," which often involved the mutual sharing of private matters, to a degree that workers sometimes felt vulnerable and exposed.[20] Much like T. S. Eliot's poet, the male call-center worker is individualized and enriched, not truly implicated, by his communicative position; while the female worker experiences a more personal, intensely felt, and at times draining participation in the process. And it is this kind of feminine connectivity that is vocationally prioritized and demanded.

In most obvious terms, the call-center worker is a medium of company expertise, but the bond she establishes may be as important as the knowledge she transmits. In other words, information exchange involves more than the exchange of information: it is not just about content or even, notwithstanding the fast pace of the information age, the expeditiousness with which it is delivered. It is about the feeling that lines the channel. Tellingly, in a fictional vision of intergalactic communications, where the human female relay does persist into the "next generation," there is an emphasis on her empathy: this is finally what a computer, for all its efficiency in transmitting data, cannot replace. Today, as at the turn of the twentieth century, we count on this power of self-extension to enable even the most functional of contacts, as we do to help us get in touch with other worlds. With our technologies that still both link and divide us, and our ghosts for which we still long, as we hope they long for us, women are still the sensitive element within networks of communication, information, and outreach to the other.

---

emotion management people bring to the workplace and exchange for a wage. As she argues, girls and women are significantly more acculturated in the practice of emotion management, and women predominate in jobs calling for emotional labor; yet this kind of work, however taxing it may be, has been "mislabeled 'natural,' a part of woman's 'being' rather than something of her own making" (167). Here we can glimpse a presumption of innate female sensitivity reminiscent of the Victorians. Hochschild notes that there is an especially high demand today for women's emotional labor in jobs entailing personal service and relations (171); see also Belt, Richardson, and Webster, "Women, Social Skill," 20–21. I would suggest that the early feminization of operating and clerical-secretarial jobs helped to pave the way for this larger trend, in that these jobs were oriented around, and set the optimal conditions for, interpersonal contact and communication as such.

20. Matthew J. Brannan, "Once More with Feeling: Ethnographic Reflections on the Mediation of Tension in a Small Team of Call Centre Workers," *Gender, Work and Organization* 12 (2005): 432, 433. As Brannan concludes, adding to the female workers' sense of vulnerability was the fact that their clients, unlike the male operators', were of a higher professional status (434–35).

❧ BIBLIOGRAPHY

Albrecht, Thomas. "Sympathy and Telepathy: The Problem of Ethics in George Eliot's *The Lifted Veil.*" *ELH* 73 (2006): 437–63.

Allen, Grant [and May Cotes]. *Kalee's Shrine.* 1886. New York: New Amsterdam Book Company, 1897.

Allen, Grant [Olive Pratt Rayner, pseud.]. *The Type-Writer Girl.* 1897. Edited by Clarissa J. Suranyi. Peterborough, Ont.: Broadview, 2004.

Anderson, Gregory. *Victorian Clerks.* Manchester: Manchester University Press, 1976.

——, ed. *The White-Blouse Revolution: Female Office Workers since 1870.* Manchester: Manchester University Press, 1988.

Arata, Stephen. *Fictions of Loss in the Victorian Fin de Siècle.* Cambridge: Cambridge University Press, 1996.

Arato, Andrew, and Eike Gebhardt, eds. *The Essential Frankfurt School Reader.* New York: Continuum, 1982.

Argyros, Ellen. *"Without Any Check of Proud Reserve": Sympathy and Its Limits in George Eliot's Novels.* New York: Peter Lang, 1999.

Armstrong, Tim. *Modernism, Technology, and the Body: A Cultural Study.* Cambridge: Cambridge University Press, 1998.

Asendorf, Christoph. *Batteries of Life: On the History of Things and Their Perception in Modernity.* Translated by Don Renau. Berkeley: University of California Press, 1993.

Ashby, Kevin. "The Centre and the Margins in 'The Lifted Veil' and Blackwood's *Edinburgh Magazine.*" *George Eliot-George Henry Lewes Studies* 24–25 (1993): 132–46.

Ashton, Rosemary. *George Eliot: A Life.* London: Allan Lane, 1996.

Atkinson, William Walker. *Practical Mind-Reading.* Chicago: Lyal Book Concern, 1908.

Baer, Ulrich. "Photography and Hysteria: Toward a Poetics of the Flash." *Yale Journal of Criticism* 7 (1994): 41–77.

Baldwin, F. G. C. *The History of the Telephone in the United Kingdom.* London: Chapman and Hall, 1938.

Balfour, Jean. "The 'Palm Sunday' Case: New Light on an Old Love Story." *Proceedings of the Society for Psychical Research* 52 (1960): 79–267.

Banta, Martha. *Henry James and the Occult: The Great Extension.* Bloomington: Indiana University Press, 1972.

Barrett, R. T. "The Changing Years as Seen from the Switchboard." *Bell Telephone Quarterly* 14 (1935): 41–295.

Bauer, Dale M., and Andrew Lakritz. "Language, Class, and Sexuality in Henry James's 'In the Cage.'" *New Orleans Review* 14, no. 3 (1987): 61–69.

Bauer, Helen Pike. *Rudyard Kipling: A Study of the Short Fiction*. New York: Twayne, 1994.

Beer, Gillian. "Myth and the Single Consciousness: *Middlemarch* and *The Lifted Veil*." In *This Particular Web: Essays on* Middlemarch, edited by Ian Adam, 91–115. Toronto: University of Toronto Press, 1975.

——. "'Wireless': Popular Physics, Radio and Modernism." In *Cultural Babbage: Technology, Time and Invention,* edited by Francis Spufford and Jenny Uglow, 149–66. London: Faber and Faber, 1996.

"Behind the Scenes at 'Central.'" *Booklovers Magazine* 2 (1903): 390–401.

Belt, Vicki, Ranald Richardson, and Juliet Webster. "Women, Social Skill and Interactive Service Work in Telephone Call Centres." *New Technology, Work and Employment* 17 (2002): 20–34.

Benjamin, Walter. "On Some Motifs in Baudelaire." In *Illuminations: Essays and Reflections,* edited by Hannah Arendt, translated by Harry Zohn, 155–200. New York: Schocken, 1968.

——. "The Work of Art in the Age of Mechanical Reproduction." In *Illuminations: Essays and Reflections,* edited by Hannah Arendt, translated by Harry Zohn, 217–51. New York: Schocken, 1968.

Blackwood, Algernon. "If the Cap Fits—." 1914. In *Ten Minute Stories,* 219–28. Freeport, NY: Books for Libraries, 1969.

——. "The Transfer." 1912. In *Best Ghost Stories of Algernon Blackwood,* compiled and introduced by E.F. Bleiler, 228–39. New York: Dover, 1973.

——. "You *May* Telephone from Here." 1914. In *Ten Minute Stories,* 170–78. Freeport, NY: Books for Libraries, 1969.

Brandon, Ruth. *The Spiritualists: The Passion for the Occult in the Nineteenth and Twentieth Centuries.* New York: Knopf, 1983.

Brannan, Matthew J. "Once More with Feeling: Ethnographic Reflections on the Mediation of Tension in a Small Team of Call Centre Workers." *Gender, Work and Organization* 12 (2005): 420–39.

Brantlinger, Patrick. *The Reading Lesson: The Threat of Mass Literacy in Nineteenth-Century British Fiction.* Bloomington: Indiana University Press, 1998.

——. *Rule of Darkness: British Literature and Imperialism, 1830–1914.* Ithaca: Cornell University Press, 1988.

Braude, Ann. *Radical Spirits: Spiritualism and Women's Rights in Nineteenth-Century America.* Boston: Beacon, 1989.

Brooks, Douglas A. "Technology, Embodiment, and Secretarial Mediation." In *Literary Secretaries/Secretarial Culture,* edited by Leah Price and Pamela Thurschwell, 129–50. Aldershot, UK: Ashgate, 2005.

Brown, Slater. *The Heyday of Spiritualism.* New York: Hawthorn Books, 1970.

Browne, Christopher. *Getting the Message: The Story of the British Post Office.* Dover, NH: Alan Sutton, 1993.

Browning, Robert. "Mr. Sludge, 'The Medium.'" 1864. In *The Poems,* edited by John Pettigrew, vol. 1, 821–60. London: Penguin Books, 1981.

Bull, Malcolm. "Mastery and Slavery in 'The Lifted Veil.'" *Essays in Criticism* 48 (1998): 244–61.

Butler, Elizabeth Beardsley. *Women and the Trades.* New York: Russell Sage Foundation, 1909.

Butters, Roger. *First Person Singular: A Review of the Life and Work of Mr. Sherlock Holmes, the World's First Consulting Detective, and His Friend and Colleague, Dr. John H. Watson.* New York: Vantage, 1984.

Cameron, Sharon. *Thinking in Henry James.* Chicago: University of Chicago Press, 1989.

Carey, James. "Technology and Ideology: The Case of the Telegraph." *Prospects: An Annual Journal of American Cultural Studies* 8 (1983): 303–25.

Carroll, Bret E. *Spiritualism in Antebellum America.* Bloomington: Indiana University Press, 1997.

Case, Alison. "Tasting the Original Apple: Gender and the Struggle for Narrative Authority in *Dracula.*" *Narrative* 1 (1993): 222–43.

Casey, Janet Galligani. "Marie Corelli and Fin de Siècle Feminism." *English Literature in Transition, 1880–1920* 35 (1992): 162–78.

CBS Corporation. "About Ghost Whisperer." CBS Shows: Ghost Whisperer. http://www.cbs.com/primetime/ghost_whisperer/about/.

Chéroux, Clément, Pierre Apraxine, Andreas Fischer, Denis Canguilhem, and Sophie Schmit. *The Perfect Medium: Photography and the Occult.* New Haven: Yale University Press, 2005.

Chicago Telephone Company. *Operators' School: First Lessons in Telephone Operating.* Chicago: Chicago Telephone Company, 1910.

Clayton, Jay. *Charles Dickens in Cyberspace: The Afterlife of the Nineteenth Century in Postmodern Culture.* Oxford: Oxford University Press, 2003.

Cohn, Samuel. *The Process of Occupational Sex-Typing: The Feminization of Clerical Labor in Great Britain.* Philadelphia: Temple University Press, 1985.

Connor, Steven. "The Machine in the Ghost: Spiritualism, Technology and the 'Direct Voice.'" In *Ghosts: Deconstruction, Psychoanalysis, History,* edited by Peter Buse and Andrew Stott, 203–25. New York: St. Martin's Press, 1999.

Corelli, Marie. *Free Opinions Freely Expressed on Certain Phases of Modern Social Life and Conduct.* London: Constable, 1905.

——. "My First Book." *Idler* 4 (1893): 239–52.

——. *A Romance of Two Worlds.* 1886. Alhambra, CA: Borden, 1986.

——. *The Sorrows of Satan.* 1895. New York: A. L. Burt, 1923.

——. *The Soul of Lilith.* 1892. London: Methuen, 1905.

Crabtree, Adam. *From Mesmer to Freud: Magnetic Sleep and the Roots of Psychological Healing.* New Haven: Yale University Press, 1993.

Crookes, William. "Some Possibilities of Electricity." *Fortnightly Review* 51 (1892): 173–81.

Czitrom, Daniel J. *Media and the American Mind: From Morse to McLuhan.* Chapel Hill: University of North Carolina Press, 1982.

Davies, Margery W. *Woman's Place Is at the Typewriter: Office Work and Office Workers, 1870–1930.* Philadelphia: Temple University Press, 1982.

——. "Women Clerical Workers and the Typewriter: The Writing Machine." In *Technology and Women's Voices: Keeping in Touch,* edited by Cheris Kramarae, 29–40. New York: Routledge and Kegan Paul, 1988.

Davison, Neil R. "'The Jew' as Homme/Femme-Fatale: Jewish (Art)ifice, *Trilby,* and Dreyfus." *Jewish Social Studies* 8 (2002): 73–111.

De Graaf, Leonard. "Thomas Edison and the Origins of the Entertainment Phonograph." *NARAS Journal* 8 (1997–98): 43–69.

Dickens, Charles. "The Rapping Spirits." 1858. In *Charles Dickens' Christmas Ghost Stories,* edited by Peter Haining, 193–200. New York: St. Martin's Press, 1993.

——. "The Signal Man." 1866. In *Charles Dickens' Christmas Ghost Stories,* edited by Peter Haining, 231–41. New York: St. Martin's Press, 1993.

Dickerson, Vanessa D. *Victorian Ghosts in the Noontide: Women Writers and the Supernatural.* Columbia: University of Missouri Press, 1996.

Doyle, Arthur Conan. "A Case of Identity." In *Sherlock Holmes: The Major Stories with Contemporary Critical Essays,* edited by John A. Hodgson, 75–90. Boston: Bedford, 1994.

——. *The Parasite.* Westminster: Constable, 1894.

——. "A Scandal in Bohemia." In *Sherlock Holmes: The Major Stories with Contemporary Critical Essays,* edited by John A. Hodgson, 32–53. Boston: Bedford, 1994.

——. *The Sign of Four.* 1890. Introduced by Peter Ackroyd, with notes by Ed Glinert. London: Penguin Books, 2001.

Draaisma, Douwe. *Metaphors of Memory: A History of Ideas about the Mind.* Translated by Paul Vincent. Cambridge: Cambridge University Press, 2000.

Du Maurier, George. *The Martian.* 1897. London: Harper and Brothers, 1898.

——. *Peter Ibbetson.* New York: Harper and Brothers, 1891.

——. *Trilby.* 1894. Edited by Daniel Pick. London: Penguin Books, 1994.

Eagleton, Terry. "Power and Knowledge in 'The Lifted Veil.'" *Literature and History* 9 (1983): 52–61.

Eliot, George. "The Lifted Veil." In *The Book of Fantastic Literature by Women, 1800–World War II,* edited and introduced by A. Susan Williams, 57–96. New York: Caroll and Graf, 1992.

Eliot, T. S. "Tradition and the Individual Talent." In *Selected Prose of T. S. Eliot,* edited by Frank Kermode, 37–44. London: Faber and Faber, 1975.

——. *The Waste Land.* In *The Longman Anthology of British Literature,* edited by David Damrosch and Kevin J. H. Dettmar. 3rd ed., vol. 2, 2518–31. New York: Longman, 2006.

Ellenberger, Henri. *The Discovery of the Unconscious: The History and Evolution of Dynamic Psychiatry.* New York: Basic Books, 1970.

Ellis, Sarah Stickney. *The Women of England: Their Social Duties and Domestic Habits.* London: Fisher, [1839].

Ermarth, Elizabeth Deeds. "George Eliot's Conception of Sympathy." *Nineteenth-Century Fiction* 40 (1985): 23–42.

Ettinger, Blanche. *Opportunities in Administrative Assistant Careers.* New York: McGraw-Hill, 2007.

Federal Trade Commission. "FTC Charges 'Miss Cleo' Promoters with Deceptive Advertising, Billing and Collection Practices." http://www.ftc.gov/opa/2002/02/accessresource.htm.

Federico, Annette R. *Idol of Suburbia: Marie Corelli and Late-Victorian Literary Culture.* Charlottesville: University Press of Virginia, 2000.

Felski, Rita. *The Gender of Modernity.* Cambridge, MA: Harvard University Press, 1995.

Fernandez, Roberto M., and M. Lourdes Sosa. "Gendering the Job: Networks and Recruitment at a Call Center." *American Journal of Sociology* 111 (2005): 859–904.

Fischer, Claude S. *America Calling: A Social History of the Telephone to 1940*. Berkeley: University of California Press, 1992.

Fleissner, Jennifer. "Dictation Anxiety: The Stenographer's Stake in *Dracula.*" *Nineteenth-Century Contexts* 22 (2000): 417–55.

Flint, Kate. "Blood, Bodies, and *The Lifted Veil.*" *Nineteenth-Century Literature* 51 (1997): 455–73.

Frank, Felicia Miller. *The Mechanical Song: Women, Voice, and the Artificial in Nineteenth-Century French Narrative*. Stanford: Stanford University Press, 1995.

Freedman, Jonathan. *The Temple of Culture: Assimilation and Anti-Semitism in Literary Anglo-America*. Oxford: Oxford University Press, 2000.

Fuller, Robert C. *Americans and the Unconscious*. Oxford: Oxford University Press, 1986.

Gabler, Edwin. *The American Telegrapher: A Social History, 1860–1900*. New Brunswick, NJ: Rutgers University Press, 1988.

Gabler-Hover, Janet. "The Ethics of Determinism in Henry James's 'In the Cage.'" *Henry James Review* 13 (1992): 253–73.

Gallon, Tom. *The Girl Behind the Keys*. London: Hutchinson, 1903.

Garland, Charles H. "Women as Telegraphists." *Economic Journal: The Journal of the British Economic Association* 11 (1901): 251–61.

Garrett, Peter K. *Gothic Reflections: Narrative Force in Nineteenth-Century Fiction*. Ithaca: Cornell University Press, 2003.

Gauld, Alan. *The Founders of Psychical Research*. New York: Schocken, 1968.

Ghose, Saroj. "Commercial Needs and Military Necessities: The Telegraph in India." In *Technology and the Raj: Western Technology and Technical Transfers to India, 1700–1947*, edited by Roy MacLeod and Deepak Kumar, 153–76. New Delhi: Sage, 1995.

Gilbert, Elliot L. *The Good Kipling: Studies in the Short Story*. [Athens] Ohio University Press, 1970.

Gissing, George. *The Odd Women*. 1893. Introduced by Elaine Showalter. London: Penguin Books, 1983.

Gitelman, Lisa. *Always Already New: Media, History, and the Data of Culture*. Cambridge, MA: MIT Press, 2006.

——. *Scripts, Grooves, and Writing Machines: Representing Technology in the Edison Era*. Stanford: Stanford University Press, 1999.

Gracombe, Sarah. "Converting Trilby: Du Maurier on Englishness, Jewishness, and Culture." *Nineteenth-Century Literature* 58 (2003): 75–108.

Gray, B M. "Pseudoscience and George Eliot's 'The Lifted Veil.'" *Nineteenth-Century Fiction* 36 (1982): 407–23.

Green, Venus. *Race on the Line: Gender, Labor, and Technology in the Bell System, 1880–1980*. Durham: Duke University Press, 2001.

Green-Lewis, Jennifer. *Framing the Victorians: Photography and the Culture of Realism*. Ithaca: Cornell University Press, 1996.

Greenway, John L. "Seward's Folly: *Dracula* as a Critique of 'Normal Science.'" *Stanford Literature Review* 3 (1986): 213–30.

Grove, Allen W. "Röntgen's Ghosts: Photography, X-Rays, and the Victorian Imagination." *Literature and Medicine* 16 (1997): 141–73.

Gunning, Tom. "Phantom Images and Modern Manifestations: Spirit Photography, Magic Theater, Trick Films, and Photography's Uncanny." In *Fugitive Images:*

From Photography to Video, edited by Patrice Petro, 42–71. Bloomington: Indiana University Press, 1995.

Hacking, Ian. *Rewriting the Soul: Multiple Personality and the Sciences of Memory.* Princeton: Princeton University Press, 1995.

Haggard, H. Rider. *She.* 1887. Edited by Daniel Karlin. Oxford: Oxford University Press, 1991.

Haller, John S., Jr., and Robin M. Haller. *The Physician and Sexuality in Victorian America.* Urbana: University of Illinois Press, 1974.

Harrington, Anne. "Hysteria, Hypnosis, and the Lure of the Invisible: The Rise of Neo-Mesmerism in *Fin-de-Siècle* French Psychiatry." In *The Asylum and Its Psychiatry,* edited by W.F. Bynum, Roy Porter, and Michael Shepherd. Vol. 3, *Anatomy of Madness: Essays in the History of Psychiatry,* 226–46. New York: Routledge, 1988.

Hartnell, Elaine M. "Morals and Metaphysics: Marie Corelli, Religion and the Gothic." *Women's Writing* 13 (2006): 284–303.

Hayles, N. Katherine. "Escape and Constraint: Three Fictions Dream of Moving from Energy to Information." In *From Energy to Information: Representation in Science and Technology, Art, and Literature,* edited by Bruce Clarke and Linda Dalrymple Henderson, 235–54. Stanford: Stanford University Press, 2002.

Hazelgrove, Jenny. *Spiritualism and British Society Between the Wars.* Manchester: Manchester University Press, 2000.

"The Health of Telephone Operators." *Lancet* 2, no. 2 (1911): 1716.

Heaton, J. Henniker. "Imperial Telegraph System: Cabling to India and Australia." *Contemporary Review* 63 (1893): 537–49.

Helsinger, Elizabeth K., Robin Lauterbach Sheets, and William R. Veeder. *The Woman Question: Society and Literature in Britain and America, 1837–1889.* Vol. 2, *Social Issues.* Chicago: University of Chicago Press, 1983.

Hensoldt, Heinrich. "Occult Science in Thibet." *Arena* 10 (1894): 366–78.

———. "The Wonders of Hindoo Magic." *Arena* 10 (1894): 46–60.

Hertz, Neil. *George Eliot's Pulse.* Stanford: Stanford University Press, 2003.

Hochman, Barbara. *Getting at the Author: Reimagining Books and Reading in the Age of American Realism.* Amherst: University of Massachusetts Press, 2001.

Hochschild, Arlie Russell. *The Managed Heart: Commercialization of Human Feeling.* 1983. Twentieth Anniversary Edition. Berkeley: University of California Press, 2003.

Holman, Rosemary Batt, and Ursula Holtgrewe. "The Global Call Center Report: International Perspectives on Management and Employment." DigitalCommons@ILR. Cornell University ILR School. http://digitalcommons.ilr.cornell.edu/cgi/viewcontent.cgi?article=1012&context=reports.

Hope, Miss Mabel. "Evidence on Behalf of the Female Telegraphists in the Central Telegraph Office, Counter Clerks and Telegraphists in the Metropolitan Districts, Returners at Mount Pleasant, and Telephonists at G.P.O. South." London: Co-operative Printing Society, 1906.

Howells, William Dean. *The Undiscovered Country.* 1880. Boston: Houghton, Mifflin, 1896.

Hurley, Kelly. *The Gothic Body: Sexuality, Materialism, and Degeneration at the Fin de Siècle.* Cambridge: Cambridge University Press, 1996.

Hutchinson, Stuart. "James's *In the Cage:* A New Interpretation." *Studies in Short Fiction* 19 (1982): 19–25.

Huyssen, Andreas. *After the Great Divide: Modernism, Mass Culture, Postmodernism.* Bloomington: Indiana University Press, 1986.

Jaffe, Audrey. *Scenes of Sympathy: Identity and Representation in Victorian Fiction.* Ithaca: Cornell University Press, 2000.

James, Henry. *The Bostonians.* 1886. Edited by Richard Lansdown. London: Penguin Books, 2001.

———. "George Du Maurier." *Harper's New Monthly Magazine* 95 (1897): 594–609.

———. *In the Cage.* 1898. In *In the Cage and Other Tales,* edited by Morton Dauwen Zabel, 174–266. Garden City, NY: Doubleday, 1958.

———. *The Turn of the Screw.* 1898. In *The Turn of the Screw and the Aspern Papers,* edited by Anthony Curtis, 143–262. London: Penguin Books, 2003.

James, William. "The Hidden Self." *Scribner's Magazine* 7 (1890): 361–73.

Jann, Rosemary. "Saved by Science? The Mixed Messages of Stoker's *Dracula.*" *Texas Studies in Language and Literature* 31 (1989): 273–87.

Jenkins, Emily. "*Trilby:* Fads, Photographers, and 'Over-Perfect Feet.'" In *Book History,* edited by Ezra Greenspan and Jonathan Rose, vol. 1, 221–67. University Park: Pennsylvania State University Press, 1998.

Jepsen, Thomas C. *My Sisters Telegraphic: Women in the Telegraph Office, 1846–1950.* Athens: Ohio University Press, 2000.

———. "Women Telegraph Operators on the Western Frontier." *Journal of the West* 35 (1996): 72–80.

Karlin, Daniel. *Browning's Hatreds.* Oxford: Oxford University Press, 1993.

Karpenko, Lara. "Purchasing Largely: *Trilby* and the *Fin de Siècle* Reader." *Victorians Institute Journal* 34 (2006): 215–42.

Keats, John. "The Eve of St. Agnes." In *The Longman Anthology of British Literature,* edited by David Damrosch and Kevin J. H. Dettmar. 3rd ed., vol. 2, 935–46. New York: Longman, 1993.

Keep, Christopher. "Blinded by the Type: Gender and Information Technology at the Turn of the Century." *Nineteenth-Century Contexts* 23 (2001): 149–73.

———. "The Cultural Work of the Type-Writer Girl." *Victorian Studies* 40 (1997): 401–26.

Kelly, Richard. *George Du Maurier.* Boston: Twayne, 1983.

Kemp, Sandra. *Kipling's Hidden Narratives.* Oxford: Blackwell, 1988.

Kenny, Michael G. *The Passion of Ansel Bourne: Multiple Personality in American Culture.* Washington, DC: Smithsonian Institution Press, 1986.

Kerr, Richard. *Wireless Telegraphy, Popularly Explained.* London: Seeley, 1903.

Kieve, Jeffrey. *The Electric Telegraph: A Social and Economic History.* Newton Abbot, UK: David and Charles, 1973.

Kipling, Rudyard. "Wireless." 1902. In *Traffics and Discoveries,* 197–221. 1904. New York: Doubleday, Page, 1914.

Kittler, Friedrich. *Discourse Networks, 1800/1900.* Translated by Michael Metteer, with Chris Cullens. Stanford: Stanford University Press, 1990.

———. "Dracula's Legacy." Translated by William Stephen Davis. In *Literature, Media, Information Systems: Essays,* edited by John Johnston, 50–84. Amsterdam: G + B Arts International, 1997.

——. *Gramophone, Film, Typewriter.* Translated by Geoffrey Winthrop-Young and Michael Wutz. Stanford: Stanford University Press, 1999.

Kowalczyk, Richard L. "In Vanished Summertime: Marie Corelli and Popular Culture." *Journal of Popular Culture* 7 (1974): 850–63.

Kreilkamp, Ivan. *Voice and the Victorian Storyteller.* Cambridge: Cambridge University Press, 2005.

Kwolek-Folland, Angel. *Engendering Business: Men and Women in the Corporate Office, 1870–1930.* Baltimore: Johns Hopkins University Press, 1994.

Le Bon, Gustave. *The Crowd: A Study of the Popular Mind.* 1896. Atlanta: Cherokee, 1982.

Ledger, Sally. *The New Woman: Fiction and Feminism at the Fin de Siècle.* Manchester: Manchester University Press, 1997.

Leys, Ruth. *Trauma: A Genealogy.* Chicago: University of Chicago Press, 2000.

Lipartito, Kenneth. "When Women Were Switches: Technology, Work, and Gender in the Telephone Industry, 1890–1920." *American Historical Review* 99 (1994): 1075–111.

London, Bette. "Mediumship, Automatism, and Modernist Authorship." In *Gender in Modernism: New Geographies, Complex Intersections,* edited by Bonnie Kime Scott, 623–32. Urbana: University of Illinois Press, 2007.

——. "Secretary to the Stars: Mediums and the Agency of Authorship." In *Literary Secretaries/Secretarial Culture,* edited by Leah Price and Pamela Thurschwell, 91–110. Aldershot, UK: Ashgate, 2005.

——. *Writing Double: Women's Literary Partnerships.* Ithaca: Cornell University Press, 1999.

"The Lot of the Woman Telegrapher: An Open Letter." Chicago: Women's Trade Union League of Illinois, [1907?].

Lubrano, Annteresa. *The Telegraph: How Technology Innovation Caused Social Change.* New York: Garland, 1997.

Luckhurst, Roger. *The Invention of Telepathy: 1870–1901.* Oxford: Oxford University Press, 2002.

Lustig, T. J. *Henry James and the Ghostly.* Cambridge: Cambridge University Press, 1994.

Maddox, Brenda. "Women and the Switchboard." In *The Social Impact of the Telephone,* edited by Ithiel de Sola Pool, 262–80. Cambridge, MA: MIT Press, 1977.

Mantel, Hilary. *Beyond Black.* New York: Picador, 2005.

Marryat, Florence. *The Dead Man's Message.* 1894. New York: Arno, 1976.

Marsh, Richard. *The Beetle.* 1897. Edited by Julian Wolfreys. Peterborough, Ont.: Broadview, 2004.

Marshall, Gail. *Actresses on the Victorian Stage: Feminine Performance and the Galatea Myth.* Cambridge: Cambridge University Press, 1998.

Martin, Michèle. *"Hello, Central?": Gender, Technology, and Culture in the Formation of Telephone Systems.* Montreal: McGill-Queen's University Press, 1991.

Marvin, Carolyn. *When Old Technologies Were New: Thinking about Electric Communication in the Late Nineteenth Century.* New York: Oxford University Press, 1988.

Mason, Diane. "Latimer's Complaint: Masturbation and Monomania in George Eliot's *The Lifted Veil.*" *Women's Writing* 5 (1998): 393–403.

Masters, Brian. *Now Barrabas Was a Rotter: The Extraordinary Life of Marie Corelli.* London: Hamish Hamilton, 1978.

Mazzoni, Cristina. *Saint Hysteria: Neurosis, Mysticism, and Gender in European Culture.* Ithaca: Cornell University Press, 1996.

[McCarthy, Justin]. "Along the Wires." *Harper's New Monthly Magazine* 40 (1870): 416–21.

McDougall, W. "The Case of Sally Beauchamp." *Proceedings of the Society for Psychical Research* 19 (1907): 410–31.

McLuhan, Marshall. *Understanding Media: The Extensions of Man.* 1964. Cambridge, MA: MIT Press, 1997.

Menke, Richard. "Media in America, 1881: Garfield, Guiteau, Bell, Whitman." *Critical Inquiry* 31 (2005): 638–64.

——. *Telegraphic Realism: Victorian Fiction and Other Information Systems.* Stanford: Stanford University Press, 2008.

"Mesmerism and Hypnotism." *Quarterly Review* 171 (1890): 234–59.

Miller, Karl. *Doubles: Studies in Literary History.* Oxford: Oxford University Press, 1985.

Mitchell, Sally. *The New Girl: Girls' Culture in England, 1880-1915.* New York: Columbia University Press, 1995.

Monroe, John Warne. *Laboratories of Faith: Mesmerism, Spiritism, and Occultism in Modern France.* Ithaca: Cornell University Press, 2008.

Moody, Andrew J. "'The Harmless Pleasure of Knowing': Privacy in the Telegraph Office and Henry James's 'In the Cage.'" *Henry James Review* 16 (1995): 53–65.

Moore, R. Laurence. *In Search of White Crows: Spiritualism, Parapsychology, and American Culture.* New York: Oxford University Press, 1977.

Myers, Frederic W. H. *Human Personality and Its Survival of Bodily Death.* 2 vols. New York: Longmans, Green, 1903.

The New Office Professional's Handbook. 4th ed. Boston: Houghton Mifflin, 2001.

Nixon, Nicola. "The Reading Gaol of Henry James's *In the Cage.*" *ELH* 66 (1999): 179–201.

Noakes, Richard. "'Instruments to Lay Hold of Spirits': Technologizing the Bodies of Victorian Spiritualism." In *Bodies/Machines,* edited by Iwan Rhys Morus, 125–63. Oxford: Berg, 2002.

——. "Spiritualism, Science and the Supernatural in Mid-Victorian Britain." In *The Victorian Supernatural,* edited by Nicola Bown, Carolyn Burdett, and Pamela Thurschwell, 23–43. Cambridge: Cambridge University Press, 2004.

——. "Telegraphy Is an Occult Art: Cromwell Fleetwood Varley and the Diffusion of Electricity to the Other World." *British Journal of the History of Science* 32 (1999): 421–59.

Norwood, Stephen H. *Labor's Flaming Youth: Telephone Operators and Worker Militancy, 1878-1923.* Urbana: University of Illinois Press, 1990.

Novak, Daniel A. *Realism, Photography, and Nineteenth-Century Fiction.* New York: Cambridge University Press, 2008.

Nye, David E. *Electrifying America: Social Meanings of a New Technology, 1880–1940.* Cambridge, MA: MIT Press, 1990.

Oppenheim, Janet. *The Other World: Spiritualism and Psychical Research in England, 1850–1914.* Cambridge: Cambridge University Press, 1985.

——. *"Shattered Nerves": Doctors, Patients, and Depression in Victorian England*. New York: Oxford University Press, 1991.

Ormond, Leonée. *George Du Maurier*. Pittsburgh: University of Pittsburgh Press, 1969.

Otis, Laura. *Networking: Communicating with Bodies and Machines in the Nineteenth Century*. Ann Arbor: University of Michigan Press, 2001.

Owen, Alex. *The Darkened Room: Women, Power, and Spiritualism in Late Victorian England*. Philadelphia: University of Pennsylvania Press, 1990.

——. *The Place of Enchantment: British Occultism and the Culture of the Modern*. Chicago: University of Chicago Press, 2004.

Peters, John Durham. *Speaking into the Air: A History of the Idea of Communication*. Chicago: University of Chicago Press, 1999.

Phelan, James. *Living to Tell about It: A Rhetoric and Ethics of Character Narration*. Ithaca: Cornell University Press, 2005.

Pick, Daniel. *Svengali's Web: The Alien Enchanter in Modern Culture*. New Haven: Yale University Press, 2000.

Picker, John. *Victorian Soundscapes*. Oxford: Oxford University Press, 2003.

Pool, Ithiel de Sola. *Forecasting the Telephone: A Retrospective Technology Assessment of the Telephone*. Norwood, NJ: Ablex, 1983.

Prince, Morton. "The Development and Genealogy of the Misses Beauchamp: A Preliminary Report of a Case of Multiple Personality." *Proceedings of the Society for Psychical Research* 15 (1901): 466–83.

——. *The Dissociation of a Personality: A Biographical Study in Abnormal Psychology*. 1905. New York: Longmans, Green, 1906.

Purcell, L. Edward. "Trilby and Trilby-Mania, The Beginning of the Bestseller System." *Journal of Popular Culture* 11 (1977): 62–76.

Rainey, Lawrence. "Taking Dictation: Collage Poetics, Pathology, and Politics." *Modernism/Modernity* 5, no. 2 (1998): 123–53.

Randall, Don. "Autumn 1857: The Making of the Indian 'Mutiny.'" *Victorian Literature and Culture* 31 (2003): 3–17.

Ransom, Teresa. *The Mysterious Miss Marie Corelli: Queen of Victorian Bestsellers*. Phoenix Mill, UK: Sutton, 1999.

Review of *A Romance of Two Worlds*, by Marie Corelli, and *The Christian*, by Hall Caine. *Quarterly Review* 188 (1898): 306–37.

Richards, Thomas. *The Imperial Archive: Knowledge and the Fantasy of Empire*. London: Verso, 1993.

Roberts, Adam. "Browning, the Dramatic Monologue and the Resuscitation of the Dead." In *The Victorian Supernatural*, edited by Nicola Bown, Carolyn Burdett, and Pamela Thurschwell, 109–27. Cambridge: Cambridge University Press, 2004.

Ronell, Avital. *The Telephone Book: Technology, Schizophrenia, Electric Speech*. Lincoln: University of Nebraska Press, 1989.

Rosenberg, Edgar. *From Shylock to Svengali: Jewish Stereotypes in English Fiction*. Stanford: Stanford University Press, 1960.

Rosenzweig, Saul. "Sally Beauchamp's Career: A Psychoarchaeological Key to Morton Prince's Classic Case of Multiple Personality." *Genetic, Social, and General Psychology Monographs* 113, no. 1 (1987): 5–60.

Roth, Michael. "Hysterical Remembering." *Modernism/Modernity* 3, no. 2 (1996): 1–30.

Rowe, John Carlos. *The Other Henry James.* Durham: Duke University Press, 1998.

Royle, Nicholas. *Telepathy and Literature: Essays on the Reading Mind.* Oxford: Blackwell, 1991.

Russell, Bob. "Call Centres: A Decade of Research." *International Journal of Management Reviews* 10 (2008): 195–219.

Russo, Mary. *The Female Grotesque: Risk, Excess, and Modernity.* New York: Routledge, 1995.

Salmon, Richard. *Henry James and the Culture of Publicity.* Cambridge: Cambridge University Press, 1997.

Savoy, Eric. "'In the Cage' and the Queer Effects of Gay History." *Novel* 28 (1995): 284–307.

Sayers, Dorothy. *Strong Poison.* 1930. New York: HarperCollins, 1995.

Sconce, Jeffrey. *Haunted Media: Electronic Presence from Telegraphy to Television.* Durham: Duke University Press, 2000.

Seed, David. "The Narrative Method of *Dracula.*" *Nineteenth-Century Fiction* 40 (1985): 61–75.

Seltzer, Mark. *Bodies and Machines.* New York: Routledge, 1992.

Seymour-Smith, Martin. *Rudyard Kipling.* New York: St. Martin's Press, 1989.

Shiach, Morag. *Modernism, Labour and Selfhood in British Literature and Culture, 1890–1930.* Cambridge: Cambridge University Press, 2004.

Showalter, Elaine. Introduction to *Trilby,* by George Du Maurier, vii–xxi. Oxford: Oxford University Press, 1998.

———. *Sexual Anarchy: Gender and Culture at the Fin de Siècle.* New York: Viking, 1990.

Shuttleworth, Sally. Introduction to *The Lifted Veil and Brother Jacob,* by George Eliot, xi–l. London: Penguin Books, 2001.

Sidis, Boris. *The Psychology of Suggestion: A Research into the Subconscious Nature of Man and Society.* New York: Appleton, 1898.

Siebers, Alisha. "Marie Corelli's Magnetic Revitalizing Power." In *Victorian Literary Mesmerism,* edited by Martin Willis and Catherine Wynne, 183–202. New York: Rodopi, 2006.

Siegert, Bernhard. *Relays: Literature as an Epoch of the Postal System.* Translated by Kevin Repp. Stanford: Stanford University Press, 1999.

Skultans, Vieda. "Mediums, Controls, and Eminent Men." In *Women's Religious Experience: Cross-Cultural Perspectives,* edited by Pat Holden, 15–26. London: Croom Helm, 1983.

Small, Helen. Introduction to *The Lifted Veil/Brother Jacob,* by George Eliot, ix–xxxviii. Oxford: Oxford University Press, 1999.

Sollors, Werner. "Dr. Benjamin Franklin's Celestial Telegraph, or Indian Blessings to Gas-Lit American Drawing Rooms." *American Quarterly* 35 (1983): 458–80.

Srole, Carole. "'A Blessing to Mankind, and Especially to Womankind': The Typewriter and the Feminization of Clerical Work, Boston, 1860–1920." In *Women, Work and Technology: Transformations,* edited by Barbara Drygulski Wright et al., 84–100. Ann Arbor: University of Michigan Press, 1987.

Standage, Tom. *The Victorian Internet: The Remarkable Story of the Telegraph and the Nineteenth Century's On-line Pioneers.* New York: Walker, 1998.

Starr, Paul. *The Creation of the Media: Political Origins of Modern Communications.* New York: Basic Books, 2004.

Stead, W.T. *After Death: A Personal Narrative.* 1897. London: Review of Reviews, 1914.

———. *Real Ghost Stories.* 1891. Edited by Estelle W. Stead. New York: Doran, 1921.

Sterne, Jonathan. *The Audible Past: Cultural Origins of Sound Reproduction.* Durham: Duke University Press, 2003.

Stevens, Hugh. "Queer Henry *In the Cage.*" In *The Cambridge Companion to Henry James,* edited by Jonathan Freedman, 120–38. Cambridge: Cambridge University Press, 1998.

Stewart, Garrett. *Dear Reader: The Conscripted Audience in Nineteenth-Century British Fiction.* Baltimore: Johns Hopkins University Press, 1996.

Stiles, Anne. "Cerebral Automatism, the Brain, and the Soul in Bram Stoker's *Dracula.*" *Journal of the History of the Neurosciences* 15 (2006): 131–52.

———. "Nervous Electricity and the Neuron Doctrine in Marie Corelli's Fiction." Unpublished manuscript.

Stoker, Bram. *Dracula.* 1897. Edited by Maud Ellmann. Oxford: Oxford University Press, 1996.

Stubbs, Katherine. "Telegraphy's Corporeal Fictions." In *New Media, 1740–1915,* edited by Lisa Gitelman and Geoffrey B. Pingree, 91–111. Cambridge, MA: MIT Press, 2003.

Swann, Charles. "Déjà Vu: Déjà Lu: 'The Lifted Veil' as an Experiment in Art." *Literature and History* 5 (1979): 40–57, 86.

Sword, Helen. *Ghostwriting Modernism.* Ithaca: Cornell University Press, 2002.

Symons, Arthur. "Villiers de l'Isle-Adam." *Fortnightly Review* 66 (1899): 197–204.

Taylor, Ina. *Victorian Sisters.* London: Weidenfeld and Nicolson, 1987.

Thomas, Ronald R. *Detective Fiction and the Rise of Forensic Science.* Cambridge: Cambridge University Press, 1999.

Thurschwell, Pamela. *Literature, Technology and Magical Thinking, 1880–1920.* Cambridge: Cambridge University Press, 2001.

———. "Supple Minds and Automatic Hands: Secretarial Agency in Early Twentieth-Century Literature." *Forum for Modern Language Studies* 37 (2001): 155–68.

Tompkins, J. M. S. *The Art of Rudyard Kipling.* London: Methuen, 1959.

Trilbyana: The Rise and Progress of a Popular Novel. New York: The Critic, 1895.

Trollope, Anthony. "The Telegraph Girl." In *The Complete Short Stories,* edited by Betty Jane Slemp Breyer, vol. 4, 69–105. Fort Worth: Texas Christian University Press, 1982.

———. "The Young Women at the London Telegraph Office." *Good Words* 18 (1877): 377–84.

Trowbridge, John. "Wireless Telegraphy." *Appleton's Popular Science Monthly* 56 (1899): 59–73.

Twain, Mark. *A Connecticut Yankee in King Arthur's Court.* 1889. Edited by M. Thomas Inge. Oxford: Oxford University Press, 1997.

"Uhura as the Barbie Doll Moesha of the 1960's." *Allscifi.com.* http://www.allscifi.com/Board.asp?BoardID=196.

U.S. Department of Labor. "Secretaries and Administrative Assistants." U.S. Bureau of Labor Statistics. http://www.bls.gov/oco/ocos151.htm.

Villiers de l'Isle Adam, Auguste, Comte de. *Tomorrow's Eve.* Translated by Robert Martin Adams. Urbana: University of Illinois Press, 1982. Translation of *L'Eve future.* 1886.

Vrettos, Athena. *Somatic Fictions: Imagining Illness in Victorian Culture.* Stanford: Stanford University Press, 1995.

Walton, Patricia. *The Disruption of the Feminine in Henry James.* Toronto: University of Toronto Press, 1992.

Warhol, Robyn R. *Gendered Interventions: Narrative Discourse in the Victorian Novel.* New Brunswick, NJ: Rutgers University Press, 1989.

Warner, Marina. *Phantasmagoria: Spirit Visions, Metaphors, and Media into the Twenty-first Century.* Oxford: Oxford University Press, 2006.

Weightman, Gavin. *Signor Marconi's Magic Box: The Most Remarkable Invention of the 19th Century & the Amateur Inventor Whose Genius Sparked a Revolution.* Cambridge, MA: Da Capo, 2003.

Weliver, Phyllis. "Music, Crowd Control and the Female Performer in *Trilby.*" In *The Idea of Music in Victorian Fiction,* edited by Sophie Fuller and Nicky Losseff, 57–80. Aldershot, UK: Ashgate, 2004.

——. *Women Musicians in Victorian Fiction, 1860–1900: Representations of Music, Science and Gender in the Leisured Home.* Aldershot, UK: Ashgate, 2000.

Welsh, Alexander. *George Eliot and Blackmail.* Cambridge, MA: Harvard University Press, 1985.

Wershler-Henry, Darren. *The Iron Whim: A Fragmented History of Typewriting.* Ithaca: Cornell University Press, 2005.

Wicke, Jennifer. "Henry James's Second Wave." *Henry James Review* 10 (1989): 146–51.

——. "Vampiric Typewriting: *Dracula* and Its Media." *ELH* 59 (1992): 467–93.

Will, Barbara. "Gertrude Stein, Automatic Writing and the Mechanics of Genius." *Forum for Modern Language Studies* 37 (2001): 169–75.

Willburn, Sarah. *Possessed Victorians: Extra Spheres in Nineteenth-Century Mystical Writings.* Aldershot, UK: Ashgate, 2006.

Winter, Alison. *Mesmerized: Powers of Mind in Victorian Britain.* Chicago: University of Chicago Press, 1998.

Wise, M. Norton. "The Gender of Automata in Victorian Britain." In *Genesis Redux: Essays in the History and Philosophy of Artificial Life,* edited by Jessica Riskin, 163–95. Chicago: University of Chicago Press, 2007.

Wolfreys, Julian. *Victorian Hauntings: Spectrality, Gothic, the Uncanny and Literature.* Houndmills, UK: Palgrave, 2002.

Wood, Jane. *Passion and Pathology in Victorian Fiction.* Oxford: Oxford University Press, 2001.

Wood, T. Martin. *George Du Maurier, the Satirist of the Victorians: A Review of His Art and Personality.* London: Chatto and Windus, 1913.

Wosk, Julie. *Women and the Machine: Representations from the Spinning Wheel to the Electronic Age.* Baltimore: Johns Hopkins University Press, 2001.

Yacobi, Tamar. "Narrative Structure and Fictional Mediation." *Poetics Today* 8 (1987): 335–72.

Yeager, Jennifer A. "Opportunities and Limitations: Female Spiritual Practice in Nineteenth-Century America." *ATQ* 7 (1993): 217–28.

Young, Peter. *Person to Person: The International Impact of the Telephone.* Cambridge: Granta, 1991.

Zanger, Jules. "A Sympathetic Vibration: Dracula and the Jews." *English Literature in Transition, 1880–1920* 34 (1991): 32–44.

Zimmeck, Meta. "Jobs for the Girls: The Expansion of Clerical Work for Women, 1850–1914." In *Unequal Opportunities: Women's Employment in England 1800–1918,* edited by Angela V. John, 152–77. Oxford: Blackwell, 1986.

——. "'The Mysteries of the Typewriter': Technology and Gender in the British Civil Service, 1870–1914." In *Women Workers and Technological Change in Europe in the Nineteenth and Twentieth Centuries,* edited by Gertjan de Groot and Marlou Schrover, 67–96. London: Taylor and Francis, 1995.

INDEX

administrative assistants, 191–93
  See also secretaries
Albrecht, Thomas, 172n
Allen, Grant
  Kalee's Shrine, 73–75, 78
  The Type-Writer Girl, 7, 18n
animal magnetism. *See* mesmerism
Arato, Andrew, 115n
Argyros, Ellen, 172n
Armstrong, Tim, 14, 105n, 149n
Ashby, Kevin, 168n
automatic writing, 3, 9, 14, 62–63, 100n,
  105n, 185–86
  in cross-correspondences, 136, 177
  in *The Dissociation of a Personality* (Prince),
    150
  Kittler on, 16
  in *The Martian* (Du Maurier), 131–33
  in "Mr. Sludge, 'The Medium'" (Brown-
    ing), 162
  and privacy, 63, 66
  in *Strong Poison* (Sayers), 145
  in "Wireless" (Kipling), 175–77, 179, 181
  See also automatism
automatism, 16–17, 61–67, 137, 159
  in *The Beetle* (Marsh), 162
  in "A Case of Identity" (Doyle), 136, 138,
    141–42
  in *Dracula* (Stoker), 72–73, 78–82, 98–99
  and female medium's instrumentality,
    13–14
  and *Galaxy Quest,* 191
  in the Gothic, 61–62, 70–71, 75–76
  in *In the Cage* (James), 38
  in *Kalee's Shrine* (Allen and Cotes), 73–75
  in *The Martian* (Du Maurier), 133
  and nerves, 66, 77n
  Oriental facility with, 70–73, 78,
    80–82, 87
  in *The Parasite* (Doyle), 161–62
  in *Peter Ibbetson* (Du Maurier), 133
  and privacy, 12–13, 63–66, 71

as sensitivity's complement, 28, 66–67, 187
  in *The Soul of Lilith* (Corelli), 87, 92, 98
  and spirit mediums, 62, 66, 189, 192
  and telegraph operators, 62–63
  in *Trilby* (Du Maurier), 99–101, 103,
    114–15, 133
  and typists, 62–63, 79, 136, 138, 141–43
  See also automatic writing; machine, female
    medium as; unconscious mind

Baer, Ulrich, 127–28n
Balfour, Arthur, 4, 136
Banta, Martha, 36
Baudelaire, Charles, 105, 116
Bauer, Dale M., 40–41n, 53n
Beard, George, 149
Beauchamp, Christine L. *See* Prince, Morton:
  The Dissociation of a Personality
Beer, Gillian, 165n, 182n
Bell, Alexander Graham, 3
Bell Telephone, 5, 13n, 27, 34, 53, 64, 67n
Benjamin, Walter, 103–5, 115–17
Besant, Annie, 21n
Binet, Alfred, 158
blackmail, 13, 46–47, 50
Blackwood, Algernon
  "If the Cap Fits—" 121
  "The Transfer," 121
  "You *May* Telephone from Here," 24
Blavatsky, Helena Petrovna, 21n
Brantlinger, Patrick, 109, 151–52n
Breuer, Josef, 124n
Brooks, Douglas A., 191n
Browning, Robert
  "'Mr. Sludge, 'The Medium,'"
    42n, 160–64
Bull, Malcolm, 166n

call centers, 193–94
Cameron, Sharon, 48–49
Carroll, Lewis, 4
Case, Alison, 79n

Casey, Janet Galligani, 91n
Charcot, Jean-Martin, 76, 81, 127n, 127–28n, 149
clairvoyance
    in *L'Eve future* (Villiers), 118
    and *The Lifted Veil* (Eliot), 166, 168n, 169–70, 172
    and mesmerism, 3
    photographic models for, 121–22, 125
Clayton, Jay, 15n
clerical work, definition of, 5n
Cohn, Samuel, 5n, 6nn13–14
communication
    of essential selfhood, 15–16
    femininity of human-mediated, 64–65
    modern concept of, 24
computer, 188, 191, 194
Cook, Florence, 10
Cooke, William Fothergill, 3
Corelli, Marie, 22, 71, 75, 192
    and Christianity, 87–90, 94–95, 98
    critical reception of, 85, 93
    Free Opinions Freely Expressed on Certain Phases of Modern Social Life and Conduct, 90–92
    and inspired female artist, 92–95, 97
    and narration, 96–97, 113
    and photography, 86, 97
    as popular author, 2, 85–86
    and public image, 86, 96n, 97
    A Romance of Two Worlds, 85, 87–90, 91n, 94–97, 113, 162, 190
    and separate spheres, 90–93
    The Sorrows of Satan, 1–2, 93, 98
    The Soul of Lilith, 1–2, 86–88, 90–94, 98
    and sympathy, 96–97, 113–14
    and typists, 95–96
Cotes, May
    Kalee's Shrine, 73–75, 78
Crabtree, Adam, 123–24
Crookes, William, 4, 10
cross-correspondences, 136, 177
Crossing over with John Edward, 189
crowds, 67–68, 104–5, 134

Daguerre, Louis, 122
⌐avison, Neil R., 132–33n
⌐eneration, 102, 128–29, 132, 152
    ⌐also evolution
    ⌐e fiction, as genre, 136, 145, 159
    ⌐Charles
    ⌐pping Spirits," 9
    ⌐l Man," 9
    ⌐ssa D., 166n, 172n

dissociative identity disorder. *See* multiple personalities
Doyle, Arthur Conan
    "A Case of Identity," 136–43, 159
    occult interests of, 4, 146
    The Parasite, 161–62
    "A Scandal in Bohemia," 138–42, 144, 159
    The Sign of Four, 145–46
Draaisma, Douwe, 122n
Du Maurier, George, 97
    The Martian, 102, 131–34, 190
    Peter Ibbetson, 102, 126–31, 133–34
    response to popular audience of, 111–13
    Trilby, 22, 70–71, 99–115, 117–18, 123, 126, 128, 132–34, 162

Eagleton, Terry, 165n, 167–68, 169n
eavesdropping, 13, 33, 46–48, 60, 63–64
    See also privacy and publicity
Edison, Thomas, 3, 106
    in *L'Eve future* (Villiers), 100–101, 103, 104n, 118–20
Edward, John, 189
electricity
    and crowds, 67, 105
    eroticization of, 176, 178, 181
    feminization of, 176–78, 181
    and hypnotism, 149
    and nerves, 31, 57, 89, 100, 105, 149
    in *A Romance of Two Worlds* (Corelli), 88–89
    and spiritualism, 9–11, 16, 31, 148–49, 178, 183, 188, 190
    and telepathy, 10
    in "Wireless" (Kipling), 176–81, 187
electromagnetism. *See* electricity
Eliot, George
    The Lifted Veil, 22, 164–75, 182
Eliot, T. S.
    "Tradition and the Individual Talent," 164, 182–84, 187, 194
    The Waste Land, 186–87
Ellis, Sarah Stickney, 28–29
empathy
    of call-center workers, 193
    in *Star Trek: Next Generation,* 190, 194
Ermarth, Elizabeth, 172n
eugenics, 129, 131–33
evolution, 12
    spiritual, 102, 125, 129–30
    See also degeneration

Federico, Annette, 86n, 91, 93, 94n
Felski, Rita, 91n

feminization
    of spirit mediumship, 4
    of stenography, 5, 6n, 8
    of telephone operating, 5–8, 26–27
    of typing, 5–8
Fleissner, Jennifer, 78–79
Flint, Kate, 167n
Fowler, Clara N. *See* Prince, Morton: *The
    Dissociation of a Personality*
Fox, Kate and Margaret, 3–4, 9, 135
Frank, Felicia Miller, 14n
Franklin, Benjamin, 10–11
Freedman, Jonathan, 107n
Freud, Sigmund, 121, 123–24

Gabler, Edwin, 7n, 35n
Gabler-Hover, Janet, 36n
Galaxy Quest, 191–92
Gallon, Tom
    The Girl Behind the Keys, 7, 143–45,
        159
Galton, Francis, 129n
Garrett, Peter K., 80, 165n
Gauld, Alan, 124n
Ghost Whisperer, 189–90
Gissing, George
    The Odd Women, 18n
Gitelman, Lisa, 14, 63n, 100n, 106n13,
        106n15
Gladstone, William, 4, 85
Gothic, as genre, 21, 61–62, 70–71, 75–76,
        151–52n, 152
Gracombe, Sarah, 107
Gray, B. M., 166n
Gray, Elisha, 3
Green, Venus, 27, 67n
Green-Lewis, Jennifer, 121n
Greenway, John L., 76n, 83–84n
Gurney, Edmund, 124, 147

Haggard, Rider
    She, 151–52
Hartnell, Elaine M., 86n
Hayles, N. Katherine, 48
Hertz, Neil, 165n
Hochman, Barbara, 108–10
Hochschild, Arlie Russell, 193–94n
Holmes, Oliver Wendell, 176
Home, Daniel Dunglas, 161, 166, 173–75
Howells, William Dean
    The Undiscovered Country, 150, 152
Hurley, Kelly, 162n
Hutchinson, Stuart, 51n
Huyssen, Andreas, 105, 117n

hypnotism, 4, 59, 67, 121–24, 133–34, 146
    and crowds, 67
    cultural ambiguity of, 74–76
    in *The Dissociation of a Personality,* 137,
        147–49, 151, 157–59
    in *Dracula* (Stoker), 76, 78–83
    and electromagnetism, 149
    in *L'Eve future* (Villiers), 101, 118, 120
    in *Peter Ibbetson* (Du Maurier), 126–27
    phonograph as metaphor for, 28n, 100–101
    and sensitivity, 100
    and sexual exploitation, 158
    See also mesmerism
hysteria, 124, 127, 131n, 152
    in crowds, 104
    and hypnotism, 59, 75–76, 124, 127,
        127–28n, 146, 149
    Myers on, 102, 125

Indian Mutiny, 69
information
    versus material possession, 48
    and narrator as medium, 167
    and privacy, 13

Jaffe, Audrey, 45
James, Henry, 22, 63, 108–9
    The Bostonians, 107n
    and Du Maurier, 101, 107n, 111–12
    In the Cage, 24, 33, 35–53, 55–56, 59–60,
        114
    and motif of feminine sensitivity, 36, 51
    and motif of the occult, 36, 49
    The Turn of the Screw, 51, 131n
James, William, 4, 16, 36, 149
Janet, Pierre, 123, 127n, 149
Jepsen, Thomas, 5n, 6n, 33n

Karlin, Daniel, 161n
Karpenko, Lara, 106n
Keats, John
    "The Eve of St. Agnes," 175, 179–80
Keep, Christopher, 18n, 63, 65n, 142n
Kelly, Richard, 127n
Kenny, Michael, 149n, 154
Kingsford, Anna, 21n
Kipling, Alice ("Trix"), 177
Kipling, Rudyard
    "Wireless," 162, 175–82
Kittler, Friedrich, 14–16, 83n, 102, 119–21,
        123
Kowalczyk, Richard L., 93
Kreilkamp, Ivan, 112n
Kwolek-Folland, Angel, 28n

labor, women as, 6
Lacan, Jacques, 15–16n, 120
Lakritz, Andrew, 40–41n, 53n
Le Bon, Gustave, 67, 129
Ledger, Sally, 35n
Lewes, George Henry, 173
Leys, Ruth, 147n, 157
literature as relayed communication, 18,
    162–64
    See also narrator as medium
Lodge, Oliver, 4, 181
Lombroso, Cesare, 4, 81
London, Bette, 17n, 63n, 185, 186n
Luckhurst, Roger, 30n, 36n, 68n, 70n, 102,
    131n
Lustig, T. J., 36n

machine, female medium as, 1, 13–15, 17,
    64, 100, 106
    in "A Case of Identity" (Doyle),
        136, 138, 141–42
    in Galaxy Quest, 191
    in The Girl Behind the Keys (Gallon),
        143–44
    in The Soul of Lilith (Corelli), 1, 88, 92
    See also automatism
male occult sensitivity, emasculation
    and, 160–62, 164–65, 177–82, 184
Mantel, Hilary
    Beyond Black, 188–89
Marconi, Guglielmo, 3, 181
Marryat, Florence
    The Dead Man's Message, 66
Marsh, Richard
    The Beetle, 161–62
Marshall, Gail, 104n
Martin, Michèle, 26–27n, 64
Marvin, Carolyn, 18n
Mason, Diane, 168n
mass media, 101–02, 104, 106–7, 115–17
McCarthy, Justin
    "Along the Wires," 25–26, 30–33, 36,
        45–46, 57
McDougall, William, 152–53
McLuhan, Marshall, 108
Medium, 189
Menke, Richard, 23n, 38–39n, 165n, 167n
The Mentalist, 190n
Mesmer, Franz Anton, 3, 10, 89, 149
mesmerism, 3–4, 10, 18
    capitalism linked to, 106–7
    and crowds, 67, 105
    cultural ambiguity of, 74–76
    in Kalee's Shrine (Allen and Cotes), 74–75

and The Lifted Veil (Eliot), 166, 168n
    phonograph as metaphor for, 28n,
        100–101
    in A Romance of Two Worlds (Corelli),
        88–89
    and sensitivity, 30, 100
    and sympathy, 29
    in Trilby (Du Maurier), 99–101, 105–8,
        110n, 115, 162
    See also hypnotism
Miller, Karl, 147n
Mitchell, Sally, 6–7n, 7n
Modernism, 24n, 105n, 109, 184, 186–87
Monroe, John Warne, 124n
Moody, Andrew J., 46–47n
Morse, Samuel, 3
multiple personalities, 59, 82n, 118, 124,
    127–28, 132, 146
    See also hysteria; Prince, Morton: The Dis-
        sociation of a Personality
Myers, Frederic, 16, 147, 165
    and spiritual continuity, 4, 102, 121, 125
    and subliminal self, 75, 124–25, 146

narration
    and sympathy, 18, 97, 111–13, 132,
        172–74
    unreliable, 167–75
narrator as medium, 18
    in The Lifted Veil (Eliot), 165–67,
        174–75
    in The Martian (Du Maurier), 131
    in A Romance of Two Worlds (Corelli),
        96–97
    in Trilby (Du Maurier), 110–14
    See also literature as relayed
        communication
neo-Hermeticism, 21n
nerves
    in "Along the Wires," 30–31, 57
    and automatism, 66, 77n
    and carpal tunnel syndrome, 57–58
    in The Dissociation of a Personality (Prince),
        148–49, 155
    in Dracula (Stoker), 76–78, 83
    and electricity, 31, 57, 89, 100, 105, 149
    in In the Cage (James), 44–45
    in Kalee's Shrine (Allen and Cotes), 74
    in The Lifted Veil (Eliot), 165, 168n,
        170–71
    and mesmerism, 30, 74
    and nerve strain, 56–59, 76–78, 148–49,
        155
    and neurasthenia, 57n, 148–49

in *A Romance of Two Worlds* (Corelli),
    88–89
and sensitivity, 12, 16–17, 30–32, 56–59,
    66, 148
of spirit mediums, 16, 30, 59, 173
and sympathy, 12, 16–17, 31, 77
telegraph network as, 31–32, 69
of telegraph operators, 30–32, 44–45,
    56–58
and telepathy, 30
telephone network as, 32
of telephone operators, 32, 56–59
of typists, 58
New Woman, 18, 78, 155, 176, 182
Nixon, Nicola, 38n
Noakes, Richard, 9n, 11n
Norwood, Stephen, 34n
Novak, Daniel, 121n, 129n
Nye, David E., 176n

occult, the
    and exoticism, 152, 189
    Oriental facility with, 69–73, 78, 80–82
    relationship to knowledge, 135
    usage of term, 21n
    See also clairvoyance; mesmerism; psychi-
        cal research; psychometry; spiritual-
        ism; telepathy
Oppenheim, Janet, 4nn3–4, 57n, 59n
Orient, the
    and automatism, 70–73, 78, 80–82
    and telepathy, 69–72, 78, 81–82
Ormond, Leonée, 111
Otis, Laura, 44n, 72n, 77n
Owen, Alex, 21n, 166n, 173n

Peters, John Durham, 24n, 27n, 65n, 107n,
    182n
Phelan, James, 171–72
phonograph, 3, 110
    in *Dracula* (Stoker), 72, 77, 82
    in *L'Eve future* (Villiers), 100–101, 119
    hypnotism compared to, 28n, 100–101
    Kittler on, 120
    as mass medium, 105–6
    as office machine, 106
    in *Peter Ibbetson* (Du Maurier), 102, 126,
        128, 131
    secretary as, 28n
    and spiritualized self, 102, 119, 123
    in *Trilby* (Du Maurier), 100–101, 106
    unconscious mind as, 28n, 100–102,
        121–23, 125–26, 128, 132
phonography. *See* stenography

photography, 3
    Benjamin on, 116, 117n
    and Corelli, 86, 97
    Kittler on, 15–16n
    and occult abilities, 121–22, 125
    in *Peter Ibbetson* (Du Maurier), 126–29,
        131
    in "A Scandal in Bohemia" (Doyle),
        138–42
    in *Trilby* (Du Maurier), 104, 110n, 114,
        117–18
    unconscious mind compared to, 102,
        121–22, 125–29
    See also spirit photography
Pick, Daniel, 70, 99n, 106n
Picker, John, 100n
Piper, Leonora, 65, 152
popular media. *See* mass media
Prince, Morton, 123
    The Dissociation of a Personality, 137, 146–59
privacy and publicity, 2, 17–18, 97–98, 136
    and administrative assistants, 192–93
    and automatism, 12–13, 63–66, 71, 143
    and Corelli, 86, 96n, 97
    in *Dracula* (Stoker), 80
    in *The Girl Behind the Keys* (Gallon),
        143–44
    in *In the Cage* (James), 53, 55
    and spirit mediums, 65–66, 136
    in *Star Trek: Next Generation*, 190
    in "The Telegraph Girl" (Trollope), 53–55
    and telegraph operators, 13, 46–48,
        53–55, 60, 136
    and telephone operators, 13, 64, 136
    in *Trilby* (Du Maurier), 105–6
    and typists, 80, 136, 143–44
prostitution, 35, 39–40, 50–53, 55, 115–17
Psych, 190n
psychical research, 4, 14, 16, 65, 75
    and Henry James, 36, 131n
    overlap with psychology, 124–25, 146–47,
        152
    and Society for Psychical Research, 4, 21,
        36, 124
Psychic Readers Network, 189
psychoanalysis, 16, 120–21, 124
psychology, 4, 123–25, 132, 146–47, 158
    See also specific psychologists
psychometry, 121–22, 146
psychophysics, 15–16, 120–21

Roberts, Adam, 163n
Romanticism, 15n, 119–20
Rosenberg, Edgar, 132–33n

Rosenzweig, Saul, 149n, 158n
Roth, Michael, 127–28n
Rowe, John Carlos, 51n, 53n
Royle, Nicholas, 172n
Russo, Mary, 104n

Salmon, Richard, 107n
Savoy, Eric, 46–47n, 47
Sayers, Dorothy
    Strong Poison, 144–45, 159
science
    in *Dracula* (Stoker), 76, 78, 80–85
    in *Kalee's Shrine* (Allen and Cotes), 74–76
    versus the occult, 74–76, 78, 80–83
    See also psychology
science fiction, 190
Sconce, Jeffrey, 11, 148–49, 182n
secretaries, 16, 17n, 191–92
    in *Dracula* (Stoker), 72, 77–80, 82–83
    in *The Girl Behind the Keys* (Gallon), 143–45
    as phonographs, 28n
    and sensitivity, 16, 17n, 28, 192
    and sympathy, 16, 17n, 27–29, 192
    typist/stenographers distinguished from, 28, 33–34, 143n, 192
    See also stenographers; typists
Seltzer, Mark, 15n
sensitivity, 12, 16–17, 55–56, 59–60, 102
    and administrative assistants, 192
    in "Along the Wires" (McCarthy), 26, 30–31
    as automatism's complement, 66–67, 187
    femininity of, 30, 66, 186–87, 193–94n, 194
    in *In the Cage* (James), 36, 44–45, 51, 55
    and mesmerism, 30, 100
    as motif in Henry James, 36, 51
    and secretaries, 16, 17n, 28, 192
    and spirit mediums, 4n, 16, 17n, 29–30, 59, 62, 100, 145, 173, 181, 183–84, 188–89
    and telegraph operators, 30–32, 44–45, 56
    and telepathy, 30
    and telephone operators, 16, 27, 32, 59
    See also nerves; sympathy
Seymour-Smith, Martin, 178
Shiach, Morag, 14n
Sholes, Christopher Latham, 3
shorthand. *See* stenography
Showalter, Elaine, 114n, 154
Shuttleworth, Sally, 167n
Sidis, Boris, 108, 122–23, 125, 146–47, 158

Siebers, Alisha, 90n
Siegert, Bernhard, 15n
Skultans, Vieda, 161n
Small, Helen, 167n
spirit mediums, 8, 11, 17n, 24, 123–24, 133
    and automatism, 62, 66, 189, 192
    feminine characteristics of, 4n, 100, 184–85
    and fraud, 10, 14, 65n, 135, 160, 173–74
    nerves of, 16, 30, 59, 148–49, 173, 183
    and privacy, 13, 65–66, 136
    and sensitivity, 4n, 16, 17n, 29–30, 59, 62, 100, 145, 173, 181, 183–84, 188–89
    and sympathy, 16, 17n, 30, 43, 145, 173
    and unconscious trains of consciousness, 146
    See also Browning, Robert: "Mr. Sludge, 'The Medium'"; spiritualism; *and specific mediums*
spirit photography, 15–16n, 118, 121, 129
spiritualism, 9, 16, 24, 67, 125, 135–36, 146
    cross-correspondences, 136, 177
    in *Beyond Black* (Mantel), 188–89
    in *Crossing Over with John Edward,* 189
    in *The Dead Man's Message* (Marryat), 66
    in *The Dissociation of a Personality* (Prince), 150–52
    and electricity, 9–11, 16, 31, 148–49, 183, 188, 190
    George Eliot's disbelief in, 173–74
    in *Ghost Whisperer,* 189–90
    in *The Girl Behind the Keys* (Gallon), 145
    in *In the Cage* (James), 36–38, 42–43, 52
    and *The Lifted Veil* (Eliot), 164, 166, 173–75
    in *Medium,* 189
    origins of, 3–4, 8–9, 135
    and religious faith, 4
    in *A Romance of Two Worlds* (Corelli), 88
    and science fiction, 190
    in *Strong Poison* (Sayers), 145
    and telegraph, 9–10, 15–16n, 67, 188
    and "Tradition and the Individual Talent" (Eliot), 182–84
    and typewriter, 145
    in "Wireless" (Kipling), 175–81
    See also spirit mediums; spirit photography
Stanley, Henry, 4
Star Trek
    Enterprise, 191
    The Next Generation, 190–91, 194
    original series, 190
Stead, W. T., 9, 68n, 121–23, 125, 127
Stein, Gertrude, 186

stenography, 5, 6n, 7–8, 106
  in *Dracula* (Stoker), 72, 79, 82
  and feminization, 5, 6n, 8
  in *The Martian* (Du Maurier), 131–33
  secretarial work distinguished from, 28,
    33–34, 143n
  See also secretaries; typists
Sterne, Jonathan, 106n, 136n
Stevens, Hugh, 46–47n
Stewart, Garrett, 109
Stiles, Anne, 78n
Stoker, Bram
  Dracula, 71–73, 76–85, 97–98, 99, 152,
    162, 192
Stubbs, Katherine, 18n, 35n, 55n
Swann, Charles, 165n
Sword, Helen, 182n, 185n, 186n
Symons, Arthur, 119
sympathy, 12, 16–18, 59–60
  and administrative assistants, 192
  in "Along the Wires" (McCarthy), 25–26,
    30–33, 45–46, 57
  and antipathy, 44–45
  and Benjamin, 115–17
  and Corelli, 96–97, 113–14
  in *Dracula* (Stoker), 73, 77, 80
  and eavesdropping, 60
  and electromagnetism, 31
  femininity of, 28–29, 66
  in the Gothic, 61, 70–71
  in *In the Cage* (James), 33, 43–46, 48,
    55–56, 59–60
  in *Kalee's Shrine* (Allen and Cotes), 74
  in *The Lifted Veil* (Eliot), 172–74
  and mesmerism, 29, 30
  as motif in Henry James, 36
  and narration, 18, 97, 111–13, 132,
    172–74
  and nerves, 12, 16–17, 30–31, 74, 77
  and secretaries, 16, 17n, 27–29, 192
  in *The Soul of Lilith* (Corelli), 91
  and spirit mediums, 16, 17n, 30, 43, 145,
    173
  and telegraph, 23, 68–69
  and telegraph operators, 25–26, 30–33,
    36n, 44–46, 48, 55
  and telephone operators, 16, 27
  in *Trilby* (Du Maurier), 110, 114–16
  See also empathy; sensitivity

tape recorder, 188
telegraph, 3, 21, 25
  crowd as, 67, 105
  in *Dracula* (Stoker), 72, 82

and empire, 68–70
  Kittler on, 15n45, 15–16n46
  as means of community, 23, 68–69
  nerves as metaphor for, 31–32, 69
  and the occult, 9–10, 15–16n, 36, 67, 188
  in *A Romance of Two Worlds* (Corelli),
    88–89
  wireless, 3, 9, 175–76, 181
telegraph operators, 4–8, 11, 17–18
  and automatism, 62–63
  marital status of, 52
  nerves of, 30–32, 44–45, 56–58
  and privacy, 13, 46–48, 53–55, 60, 136,
    175
  and the occult, 36–43, 48–49, 52, 56
  and sensitivity, 30–32, 44–45, 56, 59
  and sexual virtue, 18, 34–35, 39–40,
    50–55
  social status of, 7–8, 33–45, 47–52, 56
  and sympathy, 25–26, 30–33, 44–46, 48
  women as, 5–8
  See also James, Henry: *In the Cage;* Mc-
    Carthy, Justin: "Along the Wires";
    Trollope, Anthony: "The Telegraph
    Girl"
telepathy, 4, 9–10, 65, 102, 124–25, 152
  in crowds, 67
  in *Dracula* (Stoker), 71–72, 78, 81–82
  in *In the Cage* (James), 36n, 39–41,
    48–49, 56
  in *The Lifted Veil* (Eliot), 165–69
  and mesmerism, 3
  and nervous sensitivity, 30
  Oriental talent for, 69–72, 78, 81–82
  in *Peter Ibbetson* (Du Maurier), 126
telephone, 3, 21, 74, 88
  Kittler on, 15n
  and nerves, 32
  and the occult, 8–10, 24
telephone operators, 5–8, 11, 14, 24, 190n
  boys as, 26
  and feminization, 5–8, 26–27
  as machines, 64
  nerves of, 32, 56–59
  and privacy, 13, 64, 136
  and sensitivity, 16, 27, 32
  and sexual virtue, 34–35, 53–54
  social status of, 7–8, 27, 34–35
  and sympathy, 16, 27
teletype, 5
television, 189
  See also specific series
Theosophy, 21n
Thomas, Ronald R., 141n

Thurschwell, Pamela, 23n, 36n, 46–47n, 67n, 106n, 143
Trollope, Anthony
"The Telegraph Girl," 7, 53–55
"The Young Women at the London Telegraph Office," 54n
Twain, Mark, *A Connecticut Yankee in King Arthur's Court,* 27
typewriter, 3, 7, 15, 110, 191
in *Dracula* (Stoker), 72, 82–83
and spiritualism, 145
typists, 5–8, 11, 14–15, 67
and automatism, 62–63, 79, 136, 138, 141–43
and blind method, 63, 141n, 142
and carpal tunnel syndrome, 58
in "A Case of Identity" (Doyle), 136–39, 141–43
Corelli on, 95–96
in *Dracula* (Stoker), 72, 77–80, 82–83, 162
eroticization of, 18
and feminization, 5–8
in *The Girl Behind the Keys* (Gallon), 143–45
nerves of, 58
and privacy, 136
secretaries distinguished from, 28, 33–34, 143n, 192
social status of, 7–8, 33–34
in *Strong Poison* (Sayers), 144–45
in *The Waste Land* (Eliot), 186–87
See also secretaries; stenographers

unconscious mind, 4, 75, 147
and alternate-consciousness paradigm, 123–24, 146
in *The Martian* (Du Maurier), 102, 132, 134
Myers's concept of, 75, 124–25, 146

in *Peter Ibbetson* (Du Maurier), 102, 126–30, 133–34
as photographic, 102, 121–22, 125–29
as phonographic, 28n, 100–102, 121–23, 125–26, 128, 132
spiritual capacities of, 102, 121, 125
in *Trilby* (Du Maurier), 100, 134
See also automatism; hypnotism; mesmerism

Villiers de l'Isle Adam, Auguste, Comte de
L'Eve future, 100–101, 103, 104n, 118–20, 127, 176n
Vivé, Louis, 127
Vrettos, Athena, 106n

Walton, Patricia, 38–39n
Warhol, Robyn R., 97n
Weliver, Phyllis, 105n
Welsh, Alexander, 13, 46n
Wershler-Henry, Darren, 191
Western Union, 4
Wheatstone, Charles, 3
Wicke, Jennifer, 40–41n, 72n
Wilde, Oscar
The Picture of Dorian Gray, 128
Willburn, Sarah, 70n
Williams, Lisa, 189
Winter, Alison, 75n
Wolfreys, Julian, 165n
Wood, Jane, 165n, 168n, 170–71
Wosk, Julie, 176n

Yacobi, Tamar, 174
Yeats, George, 186
Yeats, W. B., 186

Zanger, Jules, 99n
Zimmeck, Meta, 63n